Handbook of Sustainable Development

Handbook of Sustainable Development

Edited by **Kane Harlow**

New York

Published by Callisto Reference,
106 Park Avenue, Suite 200,
New York, NY 10016, USA
www.callistoreference.com

Handbook of Sustainable Development
Edited by Kane Harlow

International Standard Book Number: 978-1-63239-415-6 (Hardback)

Printed in the United States of America.

Contents

Preface

This book highlights the importance of adopting a sustainable approach to development processes. The technological development of the human civilization has generated a consumer society spreading faster than the resources of Earth allow, with the demands on energy requirements growing exponentially in the last century. Protecting the future of human race requires an improved comprehension of the environment as well as of scientific solutions, behaviors and attitudes in line with the modes of advancement that the ecosphere of Earth can withstand. Certain veterans see this as the only solution in a global deflation of the presently unsustainable exploitation of resources; however, sustainable development presents an approach that would be practical to fuse with the managerial techniques and evaluation tools for policy and decision markers at the regional planning level. Architects, policy makers, environmentalists, engineers and economists will have to work hand in hand for the purpose of ensuring that development and planning can satisfy the current requirements of society without compromising the security of future generations. This book will provide valuable approaches and practical solutions in achieving the targets of sustainable development.

This book has been the outcome of endless efforts put in by authors and researchers on various issues and topics within the field. The book is a comprehensive collection of significant researches that are addressed in a variety of chapters. It will surely enhance the knowledge of the field among readers across the globe.

It is indeed an immense pleasure to thank our researchers and authors for their efforts to submit their piece of writing before the deadlines. Finally in the end, I would like to thank my family and colleagues who have been a great source of inspiration and support.

Editor

Part 1

Sustainable Energy:
Renewable Energy and Energy Efficiency

Life Cycle Analysis of Wind Turbine

Chaouki Ghenai
Ocean and Mechanical Engineering Department, Florida Atlantic University
USA

1. Introduction

The development of cleaner and efficient energy technologies and the use of new and renewable energy sources will play an important role in the sustainable development of a future energy strategy. The promotion of renewable sources of energy and the development of cleaner and more efficient energy systems are a high priority, for security and diversification of energy supply, environmental protection, and social and economic cohesion (International Energy Agency, 2006).

Sustainable energy is to provide the energy that meets the needs of the present without compromising the ability of future generations to meet their needs. Sustainable energy has two components: renewable energy and energy efficiency. Renewable energy uses renewable sources such biomass, wind, sun, waves, tides and geothermal heat. Renewable energy systems include wind power, solar power, wave power, geothermal power, tidal power and biomass based power. Renewable energy sources, such as wind, ocean waves, solar flux and biomass, offer emissions-free production of electricity and heat. For example, geothermal energy is heat from within the earth. The heat can be recovered as steam or hot water and use it to heat buildings or generate electricity. The solar energy can be converted into other forms of energy such as heat and electricity and wind energy is mainly used to generate electricity. Biomass is organic material made from plants and animals. Burning biomass is not the only way to release its energy. Biomass can be converted to other useable forms of energy, such as methane gas or transportation fuels, such as ethanol and biodiesel (clean alternative fuels). In addition to renewable energy, sustainable energy systems also include technologies that improve energy efficiency of systems using traditional non renewable sources. Improving the efficiency of energy systems or developing cleaner and efficient energy systems will slow down the energy demand growth, make deep cut in fossil fuel use and reduce the pollutant emissions. For examples, advanced fossil-fuel technologies could significantly reduce the amount of CO_2 emitted by increasing the efficiency with which fuels are converted to electricity. Options for coal include integrated gasification combined cycle (IGCC) technology, ultra-supercritical steam cycles and pressurized fluidized bed combustion. For the transportation sector, dramatic reductions in CO_2 emissions from transport can be achieved by using available and emerging energy-saving vehicle technologies and switching to alternative fuels such as biofuels (biodiesel, ethanol). For industrial applications, making greater use of waste heat, generating electricity on-site, and putting in place more efficient processes and equipment could minimize external energy demands from industry. Advanced process control and greater reliance on biomass and biotechnologies for producing fuels, chemicals and plastics could further reduce energy use and CO2 emissions. Energy use in residential and commercial buildings can be substantially reduced with integrated building design. Insulation, new lighting technology and efficient equipment are some of the measures that can be used to cut both energy losses and heating and cooling needs. Solar technology, on-site generation of heat and power, and computerized energy management systems within and among buildings could offer further reductions in energy use and CO2 emissions for residential and commercial buildings.

This Chapter will focus on wind energy. Electric generation using wind turbines is growing very fast. Wind energy is a clean and efficient energy system but during all stages (primary materials production, manufacturing of wind turbine parts, transportation, maintenance, and disposal) of wind turbine life cycle energy was consumed and carbon dioxide CO2 can be emitted to the atmosphere. What is the dominant phase of the wind turbine life that is consuming more energy and producing more emissions? What can be done during the design process to reduce the energy consumption and carbon foot print for the wind turbine life cycle? The first part of this chapter will include a brief history about the wind energy, the fundamental concepts of wind turbine and wind turbine parts. The second part will include a life cycle analysis of wind turbine to determine the dominant phase (material, manufacturing, use, transportation, and disposal) of wind turbine life that is consuming more energy and producing more CO2 emissions.

2. Wind energy

The use of wind as an energy source begins in antiquity. Mankind was using the wind energy for sailing ships and grinding grain or pumping water. Windmills appear in Europe back in 12th century. Between the end of nineteenth and beginning of twentieth century, first electricity generation was carried out by windmills with 12 KW. Horizontal-axis windmills were an integral part of the rural economy, but it fell into disuse with the advent of cheap fossil-fuelled engines and then the wide spread of rural electrification. However, in twentieth century there was an interest in using wind energy once electricity grids became available. In 1941, Smith-Putnam wind turbine with power of 1.25 MW was constructed in USA. This remarkable machine had a rotor 53 m in diameter, full-span pitch control and flapping blades to reduce the loads. Although a blade spar failed catastrophically in 1945, it remains the largest wind turbine constructed for some 40 years (Acker and Hand, 1999). International oil crisis in 1973 lead to re-utilization of renewable energy resources in the large scale and wind power was among others. The sudden increase in price of oil stimulated a number of substantial government-funded programs of research, development and demonstration. In 1987, a wind turbine with a rotor diameter of 97.5 m with a power of 2.5MW was constructed in USA. However, it has to be noted that the problems of operating very large wind turbines, in difficult wind climates were underestimated. With considerable state support, many private companies were constructing much smaller wind turbines for commercial sales. In particular, California in the mid-1980's resulted in the installation of very large number of quite small (less than 100 KW) wind turbines. Being smaller they were generally easy to operate and also repair or modify. The use of wind energy was stimulated in 1973 by the increase of price of fossil-fuel and of course, the main driver of wind turbines was to generate electrical power with very low CO2 emissions to help limit the climate change. In 1997 the Commission of the European Union was calling for 12 percent of the gross energy demand of the European Union to be contributed from renewable by 2010. In the last 25 years the global wind energy had been increasing drastically and at the end of 2009 total world wind capacity reached 159,213 MW. Wind power showed a growth rate of 31.7 %, the highest rate since 2001. The trend continued that wind capacity doubles every three years. The wind sector employed 550,000 persons worldwide.

In the year 2012, the wind industry is expected for the first time to offer 1 million jobs. The USA maintained its number one position in terms of total installed capacity and China became number two in total capacity, only slightly ahead of Germany, both of them with around 26,000 Megawatt of wind capacity installed. Asia accounted for the largest share of new installations (40.4 %), followed by North America (28.4 %) and Europe fell back to the third place (27.3 %). Latin America showed encouraging growth and more than doubled its installations, mainly due to Brazil and Mexico. A total wind capacity of 203,000 Megawatt will be exceeded within the year 2010. Based on accelerated development and further improved policies, world wide energy association WWEA increases its predictions and sees a global capacity of 1,900,000 Megawatt as possible by the year 2020 (World Wide Energy Association report, 2009). The world's primary energy needs are projected to grow by 56% between 2005 and 2030, by an average annual rate of 1.8% per year (European Wind Energy Agency, 2006)

2.1 Fundamental concept of wind turbine

A wind turbine is a rotary device that extracts the energy from the wind. The mechanical energy from the wind turbine is converted to electricity (wind turbine generator). The wind turbine can rotate through a horizontal (horizontal axis wind turbine – HAWT) or vertical (VAWT) axis. Most of the modern wind turbines fall in these two basic groups: HAWT and VAWT. For the HAWT, the position of the turbine can be either upwind or downwind. For the horizontal upwind turbine, the wind hits the turbine blade before it hits the tower. For the horizontal downwind turbine, the wind hits the tower first. The basic advantages of the vertical axis wind turbine are (1) the generator and gear box can be placed on the ground and (2) no need of a tower. The disadvantages of the VAWT are: (1) the wind speeds are very low close to ground level, so although you may save a tower, the wind speeds will be very low on the lower part of the rotor, and (2) the overall efficiency of the vertical axis wind turbine is not impressive (Burton et al., 2001). The main parts of a wind turbine parts (see Figure 1) are:

- **Blades**: or airfoil designed to capture the energy from the strong and fast wind. The blades are lightweight, durable and corrosion-resistant material. The best materials are composites of fiberglass and reinforced plastic.
- **Rotor**: designed to capture the maximum surface area of wind. The rotor rotates around the generator through the low speed shaft and gear box.
- **Gear Box**: A gear box magnifies or amplifies the energy output of the rotor. The gear box is situated directly between the rotor and the generator.
- **Generator:** The generator is used to produce electricity from the rotation of the rotor. Generators come in various sizes, relative to the desired power output.
- **Nacelle:** The nacelle is an enclosure that seals and protects the generator and gear box from the other elements.

- **Tower:** The tower of the wind turbine carries the nacelle and the rotor. The towers for large wind turbines may be either tubular steel towers, lattice towers, or concrete towers. The higher the wind tower, the better the wind. Winds closer to the ground are not only slower, they are also more turbulent. Higher winds are not corrupted by obstructions on the ground and they are also steadier.

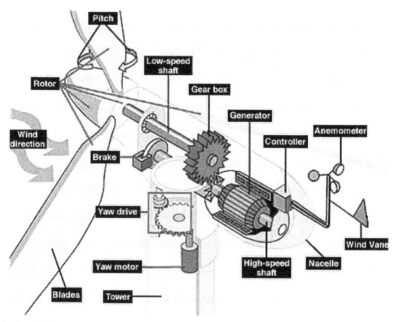

Fig. 1. Wind turbine parts

2.2 Wind turbine design

During the design of wind turbines, the strength, the dynamic behavior, and the fatigue properties of the materials and the entire assembly need to be taken into consideration. The wind turbines are built to catch the wind's kinetic energy. Modern wind turbines are not built with a lot of rotor blades. Turbines with many blades or very wide blades will be subject to very large forces, when the wind blows at high speed. The energy content of the wind varies with the third power of the wind speed. The wind turbines are built to withstand extreme winds. To limit the influence of the extreme winds and to let the turbines rotates relatively quickly it is generally prefer to build turbines with a few, long, narrow blades.

- **Fatigue Loads (forces)**: If the wind turbines are located in a very turbulent wind climate, they are subject to fluctuating winds and hence fluctuating forces. The components of the wind turbine such as rotor blades with repeated bending may develop cracks which ultimately may make the component break. When designing a wind turbine it is important to calculate in advance how the different components will vibrate, both individually, and jointly. It is also important to calculate the forces involved in each bending or stretching of a component (structural dynamics).

- **Upwind/Downwind wind turbines designs:** The upwind wind turbines have the rotor facing the wind. The basic advantage of upwind designs is that one avoids the wind shade behind the tower. By far the vast majority of wind turbines have this design. The downwind wind turbines have the rotor placed on the lee side of the tower.
- **Number of blades:** Most modern wind turbines are three-bladed designs with the rotor position maintained upwind using electrical motors in their yaw mechanism. The vast majority of the turbines sold in world markets have this design. The two-bladed wind turbine designs have the advantage of saving the cost of one rotor blade and its weight. However, they tend to have difficulty in penetrating the market, partly because they require higher rotational speed to yield the same energy output.
- **Mechanical and aerodynamics noise:** sound emissions from wind turbines may have two different origins: Mechanical noise and aerodynamic noise. The mechanical noise originates from metal components moving or knocking against each other may originate in the gearbox, in the drive train (the shafts), and in the generator of a wind turbine. Sound insulation can be useful to minimise some medium- and high-frequency noise. In general, it is important to reduce the noise problems at the source, in the structure of the machine itself. The source of the aerodynamic sound emission is when the wind hits different objects at a certain speed, it will generally start making a sound. For example, rotor blades make a slight swishing sound at relatively low wind speeds. Careful design of trailing edges and very careful handling of rotor blades while they are mounted, have become routine practice in the industry.

2.3 Wind farm

Commercial wind farms are constructed to generate electricity for sale through the electric power grid. The number of wind turbines on a wind farm can vary greatly, ranging from a single turbine to thousands. Large wind farms typically consist of multiple large turbines located in flat, open land. Small wind farms, such as those with one or two turbines, are often located on a crest or hill. The size of the turbines can vary as well, but generally they are in the range of 500 Kilowatts to several Megawatts, with 4.5 Megawatts being about the largest. Physically, they can be quite large as well, with rotor diameters ranging from 30 m to 120 m and tower heights ranging from 50 m to 100 m. The top ten wind turbine manufacturers, as measured by global market share in 2007 are listed in Table 1. Due to advances in manufacturing and design, the larger turbines are becoming more common. In general, a one Megawatt unit can produce enough electricity to meet the needs of about 100-200 average homes. A large wind farm with many turbines can produce many times that amount. However, with all commercial wind farms, the power that is generated first flows into the local electric transmission grid and does not flow directly to specific homes.

2.4 Wind turbine power

The Wind turbines work by converting the kinetic energy in the wind first into rotational kinetic energy in the turbine and then electrical energy. The wind power available for conversion mainly depends on the wind speed and the swept area of the turbine:

$$P_W = \frac{1}{2}\rho A V^3 \tag{1}$$

		Model	Power rating [kW]	Diameter [m]	Tip speed [m/s]
1	Vestas	V90	3,000	90	87
2	GE Energy	2.5XL	2,500	100	86
3	Gamesa	G90	2,000	90	90
4	Enercon	E82	2,000	82	84
5	Suzlon	S88	2,100	88	71
6	Siemens	3.6 SWT	3,600	107	73
7	Acciona	AW-119/3000	3,000	116	74.7
8	Goldwind	REpower750	750	48	58
9	Nordex	N100	2,500	99.8	78
10	Sinovel	1500 (Windtec)	1,500	70	

Table 1. Top ten wind commercial wind turbines manufactures in 2007

Where ρ is the air density (Kg/m3), A is the swept area (m2) and V the wind speed (m/s). Albert Betz (German physicist) concluded in 1919 that no wind turbine can convert more than 16/27 (59.3%) of the kinetic energy of the wind into mechanical energy turning a rotor (Betz Limit or Betz). The theoretical maximum power efficiency of any design of wind turbine is 0.59 (Hau, 2000 and Hartwanger and Horvat, 2008). No more than 59% of the energy carried by the wind can be extracted by a wind turbine. The wind turbines cannot operate at this maximum limit. The power coefficient C_p needs to be factored in equation (1) and the extractable power from the wind is given by:

$$P = \frac{1}{2} C_p \rho A V^3 \tag{2}$$

The Cp value is unique to each turbine type and is a function of wind speed that the turbine is operating in. In real world, the value of C_p is well below the Betz limit (0.59) with values of 0.35 - 0.45 for the best designed wind turbines. If we take into account the other factors in a complete wind turbine system (gearbox, bearings, generator), only 10-30% of the power of the wind is actually converted into usable electricity. The power coefficient C_p, defined as that the power extracted by rotor to power available in the wind is given by:

$$C_p = \frac{P}{\frac{1}{2}\rho A V^3} = \frac{Power\ Extracted\ by\ Rotor}{Power\ Available\ in\ the\ Wind} \tag{3}$$

3. Life cycle analysis and selections strategies for guiding design

The material life cycle is shown in Figure 2. Ore and feedstock, drawn from the earth's resources, are processed to give materials. These materials are manufactured into products that are used, and, at the end of their lives, discarded, a fraction perhaps entering a recycling loop, the rest committed to incineration or land-fill. Energy and materials are consumed at each point in this cycle (phases), with an associated penalty of CO_2, SO_x, NO_x and other emissions, heat, and gaseous, liquid and solid waste. These are assessed by the technique of life-cycle analysis (Ashby, 2005, Ashby et al., 207, Granta Design, 20090). The steps for life cycale analysis are:

1. Define the goal and scope of the assessment: Why do the assessment? What is the subject and which bit (s) of its life are assessed?
2. Compile an inventory of relevant inputs and outputs: What resources are consumed? (bill of materials) What are the emissions generated?
3. Evaluate the potential impacts associated with those inputs and outputs
4. Interpretation of the results of the inventory analysis and impact assessment phases in relation of the objectives of the study: What the result means? What is to be done about them?

The life cycle analysis studies examine energy and material flows in raw material acquisition; processing and manufacturing; distribution and storage (transport, refrigeration...); use; maintenance and repair; and recycling options (Gabi, 2008, Graedel, 1998, and Fiksel, 2009).

The eco audit or life cycle analyis and selection strategies for guiding the design are:

The first step is to develop a tool that is approximate but retains sufficient discrimination to differentiate between alternative choices. A spectrum of levels of analysis exist, ranging from a simple eco-screening against a list of banned or undesirable materials and processes to a full LCA, with overheads of time and cost.

The second step is to select a single measure of eco-stress. On one point there is some international agreement: the Kyoto Protocol committed the developed nations that signed it to progressively reduce carbon emissions, meaning CO_2 (Kyoto Protocol, 1997). At the national level the focus is more on reducing energy consumption, but since this and CO_2 production are closely related, they are nearly equivalent. Thus there is certain logic in basing design decisions on energy consumption or CO_2 generation; they carry more conviction than the use of a more obscure indicator. We shall follow this route, using energy as our measure. The third step is to separate the contributions of the phases of life because subsequent action depends on which is the dominant one. If it is that a material production, then choosing a material with low "embodied energy" is the way forward. But if it is the use phase, then choosing a material to make use less energy-intensive is the right approach, even if it has a higher embodied energy.

For selection to minimize eco-impact we must first ask: which phase of the life cycle of the product under consideration makes the largest impact on the environment? The answer guides material selection. To carry out an eco-audit we need the bill of material, shaping or manufacturing process, transportation used of the parts of the final product, the duty cycle during the use of the product, and also the eco data for the energy and CO_2 footprints of materials and manufacturing process.

The Life-Cycle Analysis has now become a vital sustainable development tool. It enables the major aspects of a product's environmental impact to be targeted, prioritization of any improvements to be made to processes, and a comparison of two products with the same function on the basis of their environmental profiles.

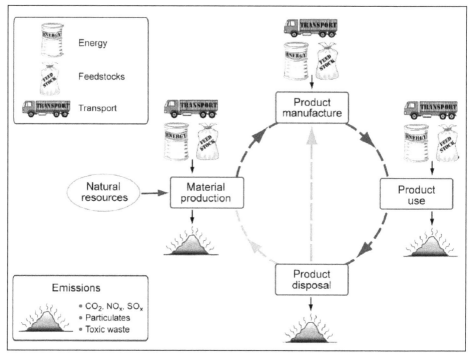

Fig. 2. Material Life cycle analysis

4. Results: Life cycle analysis of 2.0 MW wind turbine

Life cycle analysis (LCA) of 2.0 MW wind turbine is presented in this chapter. The LCA addresses the energy use and carbon foot print for the five phases (materials, manufacturing, transportation, use and disposal) through the product life cycle (Martinnez et al., 2009 and Nalukowe et al., 2006). Power generation from wind turbine is a renewable and sustainable energy but in a life cycle perspective wind turbines consumes energy resources and causes emissions during the production of raw materials, manufacturing process, its use, transportation and disposal. In order to determine the impacts of power generation using wind turbine, all components needed for the production of electricity should be include in the analysis including the tower, nacelle, rotor, foundation and transmission.

The bill of materials for a 2 MW land-based turbine (Elsam Engineering, 2004, Nordex, 2004, and Visat, 2005) is listed in Table 2. Some energy is consumed during the turbine's life (expected to be 25 years), mostly in primary materials production, manufacturing processes, and transport associated with maintenance. The energy for the transportation of small and large parts of the wind turbine and the nergy used for maintenace was calculated from information on inspection and service visits in the Vestas report (Elsam Engineering, 2004, Nordex, 2004, and Visat, 2005) and estimates of distances travelled (entered under "Static" use mode as 200 hp used for 2 hours 3 days per year). The manufacturing process for the wind turbine parst are summarized in Table 3.

Component	Material	Total Mass (kg)
Tower structure	Low carbon steel	164000.000
Tower, Cathodic Protection	Zinc alloys	203.000
Nacelle, gears	Stainless steel	19000.000
Nacelle, generator core	Cast iron, gray	9000.000
Nacelle, generator conductors	Copper	1000.000
Nacelle, transformer core	Cast iron, gray	6000.000
Nacelle, transformer conductors	Copper	2000.000
Nacelle, transformer conductors	Aluminum alloys	1700.000
Nacelle, cover	GFRP, epoxy matrix (isotropic)	4000.000
Nacelle, main shaft	Cast iron, ductile (nodular)	12000.000
Nacelle, other forged components	Stainless steel	3000.000
Nacelle, other cast components	Cast iron, ductile (nodular)	4000.000
Rotor, blades	CFRP, epoxy matrix (isotropic)	24500.000
Rotor, iron components	Cast iron, ductile (nodular)	2000.000
Rotor, spinner	GFRP, epoxy matrix (isotropic)	3000.000
Rotor, spinner	Cast iron, ductile (nodular)	2200.000
Foundations, pile & platform	Concrete	805000.000
Foundations, steel	Low carbon steel	27000.000
Transmission, conductors	Copper	254.000
Transmission, conductors	Aluminum alloys	72.000
Transmission, insulation	Polyethylene (PE)	1380.000
Total		1.091E+006

Table 2. Bill of Materials for the 2 MW Wind Turbines

Component	Manufacturing Process
Tower structure	Forging, rolling
Tower, Cathodic Protection	Casting
Nacelle, gears	Forging, rolling
Nacelle, generator core	Forging, rolling
Nacelle, generator conductors	Forging, rolling
Nacelle, transformer core	Forging, rolling
Nacelle, transformer conductors – Copper	Forging, rolling
Nacelle, transformer conductors – Aluminum	Forging, rolling
Nacelle, cover	Composite forming
Nacelle, main shaft	Casting
Nacelle, other forged components	Forging, rolling
Nacelle, other cast components	Casting
Rotor, blades	Composite forming
Rotor, iron components	Casting
Rotor, spinner	Composite forming
Rotor, spinner	Casting
Foundations, pile & platform	Construction
Foundations, steel	Forging, rolling
Transmission, conductors – Copper	Forging, rolling
Transmission, conductors – Aluminum	Forging, rolling
Transmission, insulation	Polymer extrusion

Table 3. Manufacturing Processes

The net energy demands of each phase of life are summarized in Figure 3. The life cycle analysis was performed first without recycled wind turbine materials sent to landfill). The second analysis was performed with recycled wind turbine materials (the wind turbine materials that can be recycled were sent to recycling at the end life of the wind turbine). Figure 3 and Table 4 show clearly that the dominant phase that is consuming more energy and produccing more CO_2 emisions is the material phase. More energy is consumed and high amount of CO_2 is released in the atmosphere during the primary material production of the wind turbine parts. The second dominant phase is the manufacuring process when the parts of turbine are sent to landfill at the end life of the turbine. The results also show the benefits of recycling the materials at the end life of the wind turbine. If all the materials are sent to landfill at the end of life of the wind turbine, 2.18 E+011 J of energy (1.1 % of the total energy) is needed to process these materials and 13095.71 Kg of CO_2 (0.9% increase of the total CO_2) are released to the atmosphere at the end of life of the turbine. If the material of the wind turbine are recycled, a total energy of 6.85E+012 J representing 54.8% of the total energy is recovered at the end life of the material. A net reduction of CO_2 emissions by 495917.28 Kg (55.4% of the total CO_2 emission) is obtained by recycling the wind turbine material (see Table 4).

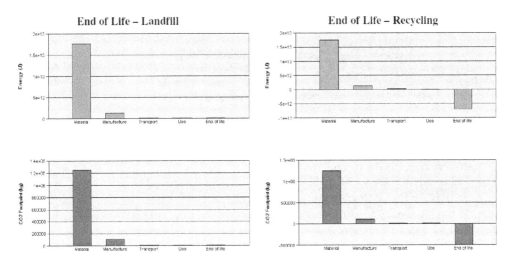

Fig. 3. Life Cycle Analysis of Wind Turbine - With and Without Wind Turbine Material Recycling

End of Life – Landfill		
Phase	Energy (J)	CO2 (kg)
Material	1.7594E+013	1.2546E+006
Manufacture	1.3593E+012	107669.7209
Transport	2.4336E+011	17278.6954
Use	1.6778E+011	11912.5577
End of life	2.1826E+011	13095.7080
Total	1.9583E+013	1.4045E+006

End of Life – Recycling		
Phase	Energy (J)	CO2 (kg)
Material	1.7594E+013	1.2546E+006
Manufacture	1.3593E+012	107669.7209
Transport	2.4336E+011	17278.6954
Use	1.6778E+011	11912.5577
End of life	-6.8512E+012	-495917.2797
Total	1.2513E+013	895503.8906

Table 4. Energy and CO2 Footprint Summary – Wind Turbine

	End of life landfill	End of life Recycling
Total Construction Energy (J)	$1.95\ 10^{13}$	$1.25\ 10^{13}$
TCE - Total Construction Energy (kWhr)	$5.41\ 10^{6}$	$3.47\ 10^{6}$
AEO - Annual Energy Output with 40% capacity factor (kWhr/year)	$7.0\ 10^{6}$	$7.0\ 10^{6}$
TE - Total Energy for the 25 years life of the turbine (kWhr)	$175\ 10^{6}$	$175\ 10^{6}$
TE/TCE - Total Energy Generated by the Turbine / Total Construction Energy	32.32	50.43
EPBT = TCE/AEO Energy Pay back Time (months)	9.27	5.94

Table 5. Construction Energy, Wind Turbine Energy Output and Energy Pay Back Time

The turbine is rated at 2 MW but it produces this power only with the right wind conditions. In a best case scenario the turbine runs at an average capacity factor of 40% giving an annual energy output of 7.0 x 10⁶ kWhr /year. The total energy generated by the turbine over a 25 year life is 175 x 10⁶ kWhr (see Table 5). The total energy generated by the turbine over 25 year life time is about 32.32 times the energy required to build and service it (5.41 10⁶ kWhr) if the turbine materials are sent to landfill at the end of life of the turbine. If the materials are recycled, the total energy generated by the turbine over 25 year life time is about 50.43 times the energy required to build and service it (3.47 10⁶ kWhr). With a wind turbine capacity factor of 40 %, the energy payback time is about 9.27 months if the wind turbine materials are sent to landfill at the end life of the turbine and is only 5.94 months if the materials are recycled. The results show clearly the benefits of recycling parts of the wind turbine at the end life of the turbine.

5. Conclusions

The development of cleaner and efficient energy technologies and the use of new and renewable energy sources will play an important role in the sustainable development of a future energy strategy. Power generation from wind turbine is a renewable and sustainable energy but in a life cycle perspective wind turbines consumes energy resources and causes emissions during the production of raw materials, manufacturing process, transportation of small and large parts of the wind turbines, maintenance, and disposal of the parts at the end life of the turbines. To determine the impacts of power generation using wind turbine, all components needed for the production of electricity should be include in the analysis including the tower, nacelle, rotor, foundation and transmission.

In eco aware wind turbine design, the materials are energy intensive with high embodies energy and carbon foot print, the material choice impacts the energy and CO_2 for the manufacturing process, the material impacts the weight of the product and its thermal and electric characteristics and the energy it consumes during the use; and the material choice also impacts the potential for recycling or energy recovery at the end of life. The eco aware wind turbine design has two-part strategy: (1) Eco Audit: quick and approximate assessment of the distribution of energy demand and carbon emission over a product's life; and (2) material selection to minimize the energy and carbon over the full life, balancing the influence of the choice over each phase of the life (selection strategies and eco informed material selection).

The results of life cycle analysis of the 2.0 MW wind turbine show the problem with the energy consumed and carbon foot print was for the material phase. More energy and more emissions are produced during the primary material production of the wind turbine parts. The manufacturing process is the second dominant phase. The energy consumption and carbon foot print are negligible for the transportation and the use phases. The results also show clearly the benefits of recycling the wind turbine parts at the end of life. The life cycle analysis of the 2.0 MW wind turbine show that 54.8% of the total energy is recovered and a net reduction of C02 emissions by 55.4% is obtained by recycling the wind turbine materials at the end of life of the wind turbine.

6. References

Acker, T.; Hand, M., (1999), "Aerodynamic Performance of the NREL Unsteady Aerodynamics Experiment (Phase IV) Twisted Rotor", AIAA-99-0045, Prepared for the 37th AIAA Aerospace Sciences Meeting and Exhibit, Reno, NV, January 11-14, p. 211-221.

Ashby, M.F., (2005) "Materials Selection in Mechanical Design", 3rd edition, Butterworth-Heinemann, Oxford, UK, Chapter 16

Ashby, M.F. Shercliff, H. and Cebon, D., (2007), "Materials: engineering, science, processing and design", Butterworth Heinemann, Oxford UK, Chapter 20.

Burton T., Sharpe D., Jenkins N. and Bossanyi E, (2001), Wind Energy Handbook, John Wiley & Sons Ltd: Chichester.

European Wind Energy Agency, VV.AA. Annual report. Technical report, EWEA, European Wind Energy Agency, 2006

Fiksel, J., Design for Envirnment, (2009), A guide to sustianble product development, McGraw Hill, ISBN 978-0-07-160556-4

Gabi, PE International, (2008), www.gabi-sofwtare.com

Graedel, T.E., (1998), Streamlined life cycle assessment, prentice Hall, ISBN 0-13-607425-1

Granta Design Limited, Cambridge, (2009) (www.grantadesign.com), CES EduPack User Guide

Hartwanger, D. and Horvat, (2008), A., 3D Modeling of a wind turbine using CFD, NAFEMS UK Conference, Cheltenham, United of Kingdom, June 10-11, 2008

Hau E, (2000), Wind turbines. Springer: Berlin.

Elsam Engineering A/S, (2004) "Life Cycle Assessment of Offshore and Onshore Sited Wind Farms", Report by Vestas Wind Systems A/S of the Danish Elsam Engineering

International Energy Agency, VV.AA, (2006), Wind energy annual report, Technical report, IEA, International Energy Agency.

Kyoto protocol, United Nations, Framework Convention on Climate Change, (1997), Document FCCC/CP 1997/7/ADD.1

Martinnez E., Sanz, F., Pellegrini, s., Jimenez e., Blanco, j., (2009), Life cycle assessment of a multi-megawatt wind turbine, Renewable Energy 34 (2009) 667–673

Nalukowe B.B., Liu, J., Damien, W., and Lukawski, T., (2006), Life Cycle Assessment of a Wind Turbine, Report 1N1800

Nordex N90 Technical Description, Nordex Energy (2004)

Vestas (2005) "Life cycle assessment of offshore and onshore sited wind turbines" Vestas Wind Systems A/S, Alsvij 21, 8900 Randus, Denmark (www.vestas.com)

Albedo Effect and Energy Efficiency of Cities

Aniceto Zaragoza Ramírez[1] and César Bartolomé Muñoz[2]
[1]Polytechnic University of Madrid
[2]Spanish Cement Association
Spain

1. Introduction

The United Nations, by means of the Intergovernmental Panel on Climate Change (IPCC), establishes in the Fourth Assessment Report, Working Group I, that warming of the climate system is unequivocal and that most of the observed increment in global average temperatures since the mid-20th century is very likely due to the observed increase in anthropogenic greenhouse gas concentrations.

European Union considers that the average surface temperature of the Earth should not be exceeded in more than 2ºC with respect to preindustrial levels in order to avoid negative consequences of global warming. With this purpose, CO_2 concentration should be kept below 450 ppmv.

The International Energy Agency (IEA) predicts an important increment of primary energy demand until 2030. The electricity generation sector expects that world's demand gets duplicated, which would mean the installation of new plants up to an additional global capacity of 5,000 GWe. This huge increment of the demand, together with other economic factors, will give fossil fuels (coal, gas and oil) a key role within the energy field.

The IPCC Third and Fourth Assessment Reports state that no individual measure by itself will be able to reduce the necessary amount of greenhouse gases emissions, but a global approach will be required. In this context, energy efficiency is considered the most relevant measure to achieve the objectives.

Despite energy efficiency shows the highest potential as a mitigation measure, the influence of the albedo of cities on global warming is not mentioned in IPCC reports, focusing on other aspects such as: thermal envelope, heating systems, co-generation and efficiency lighting systems, which are also of paramount importance, but no as powerful as albedo effect.

2. Albedo effect

2.1 Background

Looking backwards into History, we can fix the first human energy revolution when human beings abandoned the caves where they lived and set up stable settlements where new houses were built. Inside caves, the temperature was almost constant independently from the external temperature and acclimatization needs were negligible. However, new houses required new measures to keep their temperature acceptable for life.

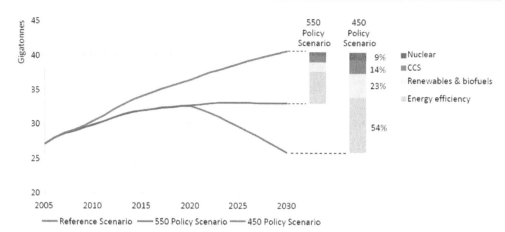

Fig. 1. Reductions in energy-related CO_2 emissions in different climate-policy scenarios (*Source: International Energy Agency, 2010*)

Fires were lighted inside houses during winter in order to protect themselves against cold. However, no air conditioning system was available for summer and our ancestors resorted to passive measures: increasing the width of walls and painting facades in light colours.

The construction of white buildings is a simple and economical bioclimatic measure to save energy, since they reflect a higher amount of solar radiation and, therefore, these buildings are cooler in summer. On the other hand, these houses are not able to absorb solar energy during winter time and they have a greater demand of heat. For this reason, we only find this kind of buildings within latitudes where solar radiation reaches a minimum value.

Fig. 2. White buildings in a town in the South of Spain

Although this value is not scientifically fixed, the experience has established that these measures are effective within latitudes under 40°, both in the north and south hemispheres, where the Earth radiation is, on average, over 225 W/m².

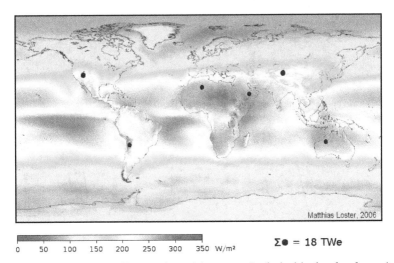

$\Sigma \bullet = 18$ TWe

Fig. 3. Solar radiation intensity (Source: http://www.ez2c.de/ml/solar_land_area)

This portion of the planet includes ¾ parts of World's population and, in consequence, these measures would have a large impact not only in developed societies, but also in developing countries from Asia or South America, where the demand of energy will become more and more important in the near future.

2.2 Earth radiation and albedo

The atmosphere of the Earth is fully transparent to visible light, but much less to infrared radiation. This is the reason why almost 58% of solar light that our planet receives reaches its surface, from which 50% is absorbed by the Earth. The rest of the radiation coming from the Sun is absorbed by the atmosphere (20% approx.), reflected by clouds (22% approx.) or reflected by Earth's surface (8% approx.).

The energy absorbed by the Earth makes it getting warm and then emitting this heat as infrared radiation, which heats up the atmosphere until it reaches a temperature at which the energy flows entering and leaving the Earth are balanced.

The presence in the Earth's atmosphere of water vapour, methane, CO_2 and other greenhouse gases, which are nearly impermeable to infrared radiation, keep the energy emitted in such a way, increasing the equilibrium temperature in comparison with the temperature in the absence of these gases. This effect is desirable, since otherwise the temperature of the Earth would be too low (-18 or -19°C) for life. However, the excessive concentration of GHG has resulted into an unusual increment of atmospheric temperature and consequently into a climate change that will modify rain distribution around the Earth, will increase the frequency of extreme atmospheric phenomena and will cause more floods, droughts and hurricanes.

The first approach from the international community to face this problem consisted of reducing the emission of GHG so that the infrared radiation emitted by the Earth is not

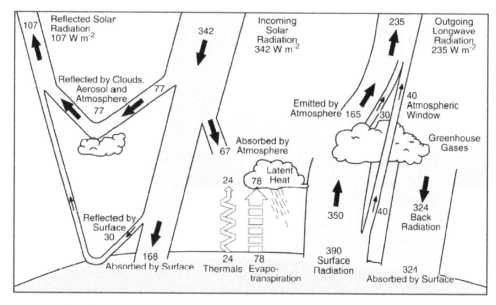

Fig. 4. Earth radiation Budget (Source: Kiehl and Trenberth, 1997)

retained in the atmosphere. Nevertheless, there exist other fields that have not been exploited yet, at least at a global scale, whose influence on global warming would be relevant. Increasing the reflectivity of the Earth's surface would decrease the absorption of solar energy, thus, emitting less infrared radiation and cooling the atmosphere. This action is cheap and means no significant changes in production processes, thus, permitting an immediate and fast deployment.

Earth's surface naturally reflects approximately 8% (not considering the poles) (Kiehl and Trenberth, 1997) of the total solar energy it receives. This reflection is conditioned by the colour of the surface: the lighter the more energy it reflects and vice versa.

The albedo or reflection coefficient is the diffuse reflectivity or reflecting power of a surface. It is defined as the ratio of reflected radiation from the surface to incident radiation upon it. A surface that absorbs all the energy it receives (black surface) has an albedo of 0, whereas a perfect reflector (white surface) has an albedo of 1.

Typical albedo of different kind of surfaces and materials are shown in table 1:

Surface	Albedo	Surface	Albedo
Fresh snow	81 - 88 %	Old snow	65 – 81 %
Ice	30 – 50 %	Rock	20 – 25 %
Woodland	5 – 15 %	Exposed soil	35 %
Oceans	5 – 10 %	Concrete	15 -25 %
Asphalt	2 – 10 %		

Table 1. Typical albedo values of different kind of surfaces and materials (*Source: European Concrete Paving Association, 2009*)

On average, the albedo of the planet is 0.35. That is to say 35% of all the solar energy is reflected while 65% is absorbed. However, it must be pointed out that polar ice, with its high albedo plays an important role in maintaining this balance. Should the polar ice melt, the average albedo of the Earth will fall because the oceans will absorb more heat than the ice.

Humans act on Earth's surface, mainly by means of construction works, usually decreasing its albedo. Pavements and building's roofs are the most exposed surfaces to solar radiation among typical construction works and they must become the objective of any measure aiming at increasing energy efficiency by means of the application of the albedo effect. In this sense, there exist at the moment construction materials that are lighter than the natural surface of the Earth. Its utilization would increase the global albedo, reducing, this way, the amount of solar energy absorbed by our planet.

Asphalt albedo ranges from about 0.05 to 0.20 (Akbari and Thayer, 2007), depending on the age and makeup of the asphalt. Its albedo typically increases somewhat as its colour fades with age. A typical concrete has an albedo of about 0.35 to 0.40 when constructed; these values can decrease to about 0.25 to 0.30 with normal usage. With the incorporation of slag or white cement, a concrete pavement can exhibit albedo readings as high as 0.70. As shown in Figure 5, concrete has a significantly higher albedo than asphalt, either new or old. In fact, concrete usually has a higher albedo than almost every other material that is typical to urban areas, including grass, trees, coloured paint, brick/stone and most roofs.

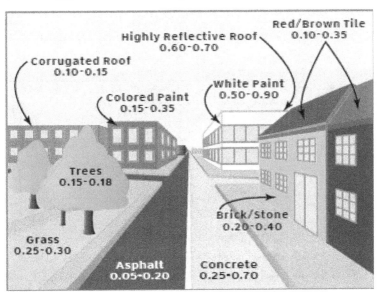

Fig. 5. Albedo ranges of various surfaces typical to urban areas *(Source: NASA, Akbari and Thayer, 2007)*

3. Influence of albedo on global warming

As previously mentioned, the albedo effect has a significant influence on climate change since approximately 50% of the solar radiation reaches the Earth's surface, where it will be either absorbed or reflected, depending on its albedo. The more radiation the Earth absorbs, the hotter it becomes, favouring global warming.

Nonetheless, this global warming effect can be slowed down by applying our knowledge, namely by providing more reflective surfaces.

The "Heat Island Group", a research group from Berkeley, California, compared the albedo effect and the influence of the concentration of atmospheric CO_2 on the net radiation power responsible for global warming. They calculated that an increase by one percent of the albedo of a surface corresponds to a reduction in radiation of 1.27 W/m^2. This reduction in radiation has the effect of slowing global warming. Their calculations indicate that delay in warming is equivalent to a reduction in CO_2 emissions of 2.5 kg per m^2 of the Earth's surface.

According to these results, the potential of this measure would be at the same level as increasing renewable energies up to 20% of the energy mix or implementing CCS technologies in power and industrial plants. Additionally, the cost of this measure would be much more economical. As construction works are continuously being carried out around the World, especially in developing countries where the deficit of infrastructures is considerable, this measure could be easily applied either by using lighter construction materials or by painting surfaces in white or other light colours. Whereas the latter means an extra cost from paint, the former could be deployed at the same price or even cheaper.

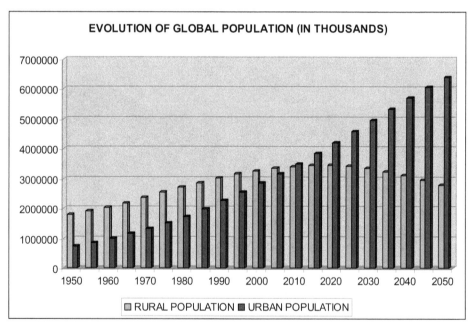

Fig. 6. Past and future evolution of worldwide population (Source: Population Division of the Department of Economic and Social Affairs of the United Nations Secretariat, World Population Prospects: The 2006 Revision and World Urbanization Prospects: The 2007 Revision)

4. The albedo and the cities

4.1 Energy demand of cities and buildings

World's population has had a steady trend to urban concentration from the last century. In 1900, the ratio of people who lived in cities represented the 10% of global population, whereas in 2007 United Nations estimated that urban population already exceeded rural people. Projections predict the same trend for the future and, therefore, present situation of cities will worsen and new problems will arise.

One of the main difficulties that cities will have to face in the future is the energy supply. Despite only representing 2 percent of the world's surface area, cities are responsible for 75 percent of the world's energy consumption. London, for example, requires a staggering 125 times its own area in resources to sustain itself (New Scientist, 2009)

London's population is around 7.4 million, so it is nowhere near megacity status yet, but according to the Tyndall Centre, it already consumes more energy than Ireland (and the same amount as Greece or Portugal) (New Scientist, 2009).

A great amount of this energy consumed by cities comes from residential uses, mainly from acclimatization of buildings.

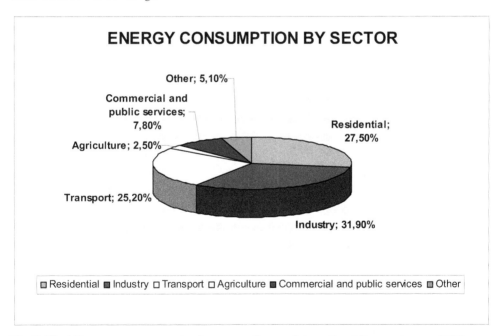

Fig. 7. World's energy consumption by sector *(Source: International Energy Agency, 2009)*

For this reason, it is essential to adopt measures aiming at reducing energy consumption of buildings within cities. Thermal insulation systems are continuously improving, electrical appliances are becoming more and more efficient and the utilization of energy-efficient light bulbs is increasing.

However, there exist measures that are not being considered and whose impact would be much higher than the impact of traditional solutions. Cement and concrete are in the core of these innovative measures, since they can increase the reflectivity of cities, helping to reduce the ambient temperature and, in consequence, the air conditioning needs.

4.2 Urban heat islands

An urban heat island (UHI) is a metropolitan area which is significantly warmer than its surrounding rural areas. This effect is more noticeable during the night than during the day and it is more apparent when the winds are weak. Seasonally, UHI is seen in both summer and winter.

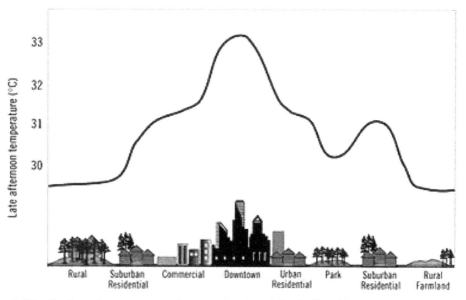

Fig. 8. Distribution of temperatures due to urban heat island effect (European Concrete Paving Association, 2009)

In the case of Madrid (Spain), two meteorological stations were chosen: one in the city centre, near Retiro Park, and another in Barajas, where the airport is placed. The distance between both stations is less than 15 kilometres, but the difference of temperatures reaches 8°C. Although Madrid is a dense city where this effect is deeper, it can also be noticed in any other city where building concentration and dark pavements produce this effect.

There are several causes for urban heat islands. The main reason is the modification of the environment by human being, introducing new materials that absorb more heat than natural ones. The solar radiation absorbed by the urban construction materials highly affects the temperature distribution inside cities. The thermo physical properties of these materials, especially the albedo and the infrared emissivity, have an important impact in their

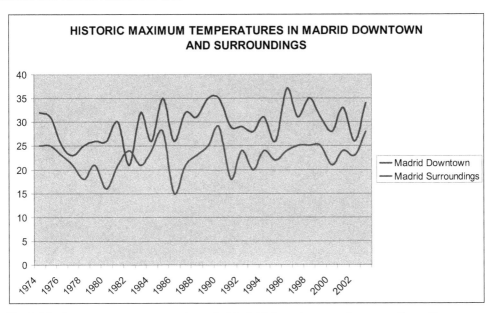

Fig. 9. Maximum average temperatures in Madrid downtown and surroundings (Source: Agencia Estatal de Meteorología, Spain)

energetic balance. Pavements are quantitatively important urban components in what is referred to the horizontal surface exposed to solar radiation (20% of the urban ground approximately) (Rose et al., 2003) and generally they have a high absorptivity and a high thermal capacity. These characteristics made significant contributions to the urban heat island effect, particularly in arid climates with high radiation levels.

This effect can be easily appreciated in a thermal photography that gives the temperature of any surface (see figure 10)

The heat absorbed during the day coming from solar radiation is emitted at night to the relatively cold night sky. The result is the increment of the average temperature within cities in approximately 5°C, which has a negative effect, mainly during summer time and in those areas where the air conditioning needs are high. The immediate result will be a dramatic rise of the demand of energy.

4.3 Strategy for cities

Increasing the albedo of the Earth's surface has a positive effect against global warming independently of the place where this increment is achieved. However, the effect of lightening the surfaces inside cities is much higher due to the indirect effect on urban heat islands and on acclimatization needs. For this reason, it must be in cities where greater efforts must be made in order to take advantage of this powerful tool.

Light roofs and facades have a direct effect on energy demand of buildings. When receiving solar radiation, these buildings reflect more light and, therefore, absorb less energy, thus,

Fig. 10. Temperatures distribution on top of surfaces built by human being (Source: "Concrete roads: a smart and sustainable choice", European Concrete Paving Association, 2009)

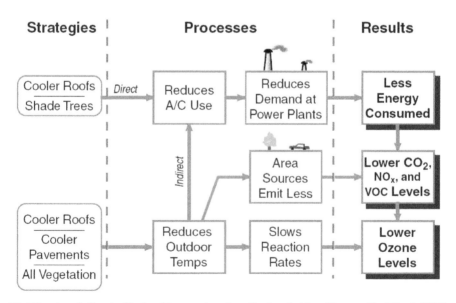

Fig. 11. Direct and direct effects of increasing the albedo of cities (*Source: H. Akbari, 2009*)

reducing the inner temperature. As a result of this, the energy demand for acclimatization purposes of the building also decreases.

Additionally, pavements can also be lighter, which means increasing the global albedo of the Earth and reflecting a higher amount of solar energy into the space.

Both measures have an indirect effect in cities. Higher albedo means less energy absorption, which helps to combat urban heat islands, reducing the ambient temperature and the energy consumption for air conditioning.

There are more advantages coming from the increase of albedo, which is not only accompanied with lower surface temperatures and lower energy use, but also with lower CO_2, NOx, VOC and ozone levels. Besides, the probability of smog formation decreases 5% for every 0.25°C fall in daily maximum temperatures above 21°C.

5. Quantification of albedo effect

5.1 Potential of global albedo effect

The most important human action on environments is, with high probability, road construction. Transport has become one of the principal activities of human being and road transport represents 85% of this activity. The consequence has been the construction of more than 30 million kilometres of roads (International Road Federation, 2006), of which 20.3 million kilometres are paved. There are other constructions that also modify the Earth's surface: railways, dams, airports, etc, but their magnitude is far from road construction. This is the reason why the quantification is only focused on roads.

The principal material used in road pavement construction is asphalt, to the point that more than 95% of roads worldwide are built with this material. As the albedo of asphalt ranges between 5 and 10% (European Concrete Paving Association, 2009), whereas the albedo of exposed soil can be fixed in 35% (European Concrete Paving Association, 2009), the result is an absolute reduction of 25% in the albedo of the surface occupied by the road. Although it has not been quantified, researches carried out in Berkeley have proved that this action have a significant impact on global warming.

Lightening asphalt or substituting it by other lighter materials (i.e. concrete) would have the contrary effect. This action would not only reduce the global albedo of the planet but it would increase it in those areas where road runs along forests or other surfaces with medium albedo values.

Assuming an average width for paved roads of 10 metres, we can estimate the influence of lightening road pavements on global warming by the application of the following equation:

$$TER = \Delta A \times ER \times ES$$

where:

ΔA = Increment of Earth albedo by asphalt-concrete substitution in roads
ER = Equivalent reduction in CO_2 emissions per m^2 per 1% increment of albedo
ES= Earth surface
ΔA can be calculated by the application of the equation:

$$\Delta A = \frac{PS}{ES} \times \Delta NA$$

where:

PS = Total paved surface

ΔNA = Difference in albedo between asphalt and concrete

The results of the calculation are shown in table 2:

PARAMETER	VALUE	SOURCE
Total length of paved roads	20,301,039 km	IRF Statistics, 2006
Average width of roads	10 m	Estimated
Total paved surface	203,104 km²	Calculated
Earth's surface	510,072,000 km²	Wikipedia http://en.wikipedia.org/wiki/Earth
Ratio paved surface/total surface	0.04 %	Calculated
Albedo of asphalt	10%	EUPAVE, 2010
Albedo of concrete	25%	EUPAVE, 2010
Difference in albedo between asphalt and concrete	15%	Calculated
Increment of Earth albedo by asphalt-concrete substitution in roads	0,006 %	Calculated
Equivalent reduction in CO_2 emissions per m² per 1% increment of albedo	2.5 kg/m²	Heat Island Group, 2009
Total equivalent reduction by asphalt-concrete substitution	7.6 GTn CO_2	Calculated

Table 2. Calculation of global albedo effect estimated in equivalent CO_2 emissions

The potential reduction of atmosphere temperature thanks to the substitution of road asphalt by lighter materials is equivalent to withdrawing from the atmosphere 7.6 G tons, which represents 25% of total emissions of CO_2 in 2010 (International Energy Agency, 2010)

Obviously, it is not feasible to lighten all paved roads around the World, since the cost of this action is unaffordable. However, this result should make the society focus on the implementation of this measure. In this sense, new roads and preservation works should become a powerful tool against global warming.

5.2 Potential of albedo effect inside cities

Quantifying the effect of an increment of the albedo inside cities is much more complicated than quantifying the global effect coming from lightening road pavements, mainly due to indirect effects. The principal benefit of increasing the albedo of cities comes from the partial elimination of heat islands, which is difficult to evaluate.

However, recent researches (Lawrence Berkeley National Laboratory in California, 2009) suggest that urban areas cover between 1.2 to 2.4% of the Earth's land mass. Paved areas in 100 of the world's largest cities cover an area of about 525 billion m². By using lighter pavements and roofs, it is possible to increase the albedo of urban surfaces in the world's top 100 cities up to 15%. The goal is to positively impact energy use, as well as to reduce smog formation, CO_2 levels, and ultimately, global warming.

Typical reflectivity of roof fabric is low, especially in residential sector. The average albedo of roofs can be estimated between 10 and 25%. It is possible to increase the albedo of roofs either by using new lighter materials or by placing a white covering on top of current roofs. In case these measures are applied, reflectivity of roofs is estimated to reach 55-60%, which means an absolute increment of 30%. Regarding pavements, long term albedo could be increased in a 10%.

A study carried out in the United States (Rose et al. 2003) estimated that the total surface occupied by roofs in urban areas is 20% and it can reach up to 25% in dense cities. The same study establishes that the percentage of paved surfaces ranges between 29 and 44%. Since the ratio of surface for vegetation purposes in American areas exceeds the world's average, it can be stated that the ratio of urban surface occupied either by roofs or by pavements is 20 and 40% respectively.

The distribution of surfaces by type inside cities is shown in table 3:

Vegetation	28 %
Roofs	20 %
Pavements	40 %
Other	12%

Table 3. Distribution of surfaces by type inside cities (Source: Rose et al. 2003, partially modified)

Assuming previous figures, an absolute increment of 0.10 in the reflectivity of albedo inside cities can be achieved (see table 4)

	Potential increment of albedo	Surface ratio	Total increment
Roof	0,30	20%	0,06
Pavement	0,10	40%	0,04
Total			0,10

Table 4. Increment of global albedo of cities due to the increment of roofs and pavements reflectivity

So far, assessing the influence of a 10% increment of albedo of cities has been extremely complicated because numerical models could only use a minimum square grid of 250 km sideway for the evaluation. Computational developments now permit reducing the size of the grid up to 50-100 km and even less for urban areas. These powerful models predict a fall of global temperature of 0,03 °C under these assumptions (H. Akbari, 2009):

1. Urban areas (2% of total Earth's surface) are perfectly white (albedo equal to 1).
2. The rest of the Earth's surface (98%) maintained its natural albedo.

In case the first assumption is changed and, instead of a perfect reflectivity, a 10% increment of the albedo is considered, the reduction of the global temperature would be 0,01 °C annually (H. Akbari, 2009). Taking into account that United Nations predict an increment of global temperature of 3°C in the next 60 years (0,05 °C per year), the conclusion is that substituting current roof and asphalt pavements by cool roofs and concrete pavements would slow down global warming by 20%.

World's current rate of CO_2 emissions is about 30 G tons/year (International Energy Agency, 2010). World's rate of CO_2 emissions averaged over next 60 years is estimated at 50 G tons/year (International Energy Agency, 2008). Hence, the 20% delay in global warming is worth 10 G ton CO_2 annually.

5.3 Economic impact

Although economic evaluation of certain measures is always difficult because of the volatility of prices, net potential savings coming from the direct effect on buildings (cooling energy savings minus heating energy penalties) have been estimated in excess of one billion Euros only in United States (H. Akbari, 2007).

Additionally, partial elimination of heat islands would make cities cooler and would improve the quality of air, thus indirectly reducing the consumption of energy for air conditioning (Taha 2002, Taha 2001, Taha et al. 2000, Rosenfeld et al. 1998; Akbari et al. 2001, Pomerantz et al. 1999). Savings of energy and better quality of air would mean a 2 billion Euros profit per year only in United States.

At the moment, the price of a CO_2 ton in European emissions market is around 13 €. This price is relatively low due to the current economic and financial crisis. However, it is estimated to rise up to 30 € in the short term. Substituting asphalt by lighter materials in world's paved roads would have the same effect as abating 7,6 G tons of CO_2 per year (calculated in this report). Global effect on cities, direct and indirect effect would be equivalent to abating 10 G tons annually (H. Akbari, 2009). The estimated cost (emissions trading) of both effects in the middle term could reach 528 billion Euros.

6. Conclusions

Results presented in this report show the social and environmental advantages of increasing the reflectivity of roofs and pavements inside cities and in interurban roads. Actually, the potential benefits of this measure are much higher than other actions that are considered as a priority by United Nations and other international organizations.

Increasing urban albedo in a 10% would allow saving up to 3 billion Euros in electricity only in the United States (H. Akbari, 2009) thanks to direct effects and also to the elimination of urban heat islands. Additionally, its influence on global temperature could slow down global warming up to 45%(calculated in this report), which would have an equivalent economic saving of more than 500 billion Euros per year during the next 60 years.

Obviously, it is not possible to reach these figures in the short term. Increasing the reflectivity of roofs means placing white materials on top of current buildings and using innovative solutions for new buildings. In the same way, increasing the albedo of pavements means covering all asphalt pavements with other construction materials, mainly concrete (whitetopping) and building new pavements of cities with concrete.

Besides, the actual effect of increasing reflectivity of construction works is not precisely defined. For this reason, it is essential to perform research and demonstration projects to deeply study and evaluate the differences between traditional and new cities with cool roofs and pavements, quantifying potential benefits and profits.

Increasing the reflectivity of urban areas and also of interurban paved roads might have the same effect as the rest of measures considered for combating global warming all together: energy efficiency of industrial processes, CCS technologies, nuclear power plants, renewable sources of energy, etc. This is the reason why increasing albedo of construction materials should be included as a mitigation measure in IPCC reports.

7. References

"Combat Global Warming", pavements4life.com, American Concrete Pavement Association, 2007.

"Energy efficiency in buildings, transforming the market", World Business Council for Sustainable Development, Switzerland, 2009.

European Directive 2002/91/CE: "Directive on the Energy Performance of Buildings"

"Energy Technology Perspectives: Scenarios & strategy to 2050", International Energy Agency, OECD/IEA, Paris, 2010

H. Akbari, S. Menon and A. Rosenfeld, "Global cooling: effect of urban albedo on global temperature", 2nd PALENC Conference and 28th AIVC Conference on Building Low energy Cooling and Advance Ventilation Technologies in the 21st Century, 2007.

H. Akbari and R. Levinson, "Status of cool roofs standards in the united States", 2nd PALENC Conference and 28th AIVC Conference on Building Low energy Cooling and Advance Ventilation Technologies in the 21st Century, 2007.

H. Akbari, "Global Cooling: increasing worldwide urban albedo to offset CO_2", Fifth Annual California Climate Change Conference, Sacramento, CA, 2009.

H. Akbari, M. Pomerantz and H. Taha, "Cool surfaces and shade trees to reduce energy use and improve air quality in urban areas", Solar Energy, vol. 70, N.3, Great Britain, 2001.

I. Ben Hamadi, P. Pouezat and A. Bastienne, "The IRF World Road Statistics", International Road Federation, Switzerland, 2006.

L. Rens, "Concrete roads: a smart and sustainable choice", European Concrete Paving Association, Brussels, 2009.

"Technology Roadmap: Energy efficient buildings: heating and cooling equipment", International Energy Agency, France, 2009.

Part 2

Sustainable Engineering and Technologies

3

Sustainability Assessment of Technologies – An Integrative Approach

Armin Grunwald

Institute for Technology Assessment and Systems Analysis (ITAS),
Karlsruhe Institute of Technology (KIT), Karlsruhe
Germany

1. Introduction

The vision of sustainable development must by definition include both long-term considerations and the global dimension (World Commission on Environment and Development, 1987; Grunwald & Kopfmüller, 2006). Pursuing this vision implies that societal processes and structures should be re-orientated so as to ensure that the needs of future generations are taken into account and to enable current generations in the southern and northern hemispheres to develop in a manner that observes the issues of equity and participation. Since a feature inherent in the Leitbild of sustainable development is consideration of strategies for shaping current and future society according to its normative content, *guidance* is necessary and the ultimate aim of sustainability analyses, reflections, deliberations, and assessments. The latter should result, in the last consequence, in *knowledge for action*, and this knowledge should motivate, empower, and support "real" action, decision-making, and planning (von Schomberg, 2002).

Technology is of major importance for sustainable development (see Sec. 2). On the one hand, technology determines to a large extent the demand for raw materials and energy, needs for transport and infrastructure, mass flows of materials, emissions as well as amount and composition of waste. Technology is, on the other side, also a key factor of the innovation system and influences prosperity, consumption patterns, lifestyles, social relations, and cultural developments. The development, production, use, and disposal of technical products and systems have impacts on all dimensions of sustainable development. Therefore, a sustainability assessment of the development, use, and disposal of technologies is required as an element of comprehensive sustainability strategies.

Technology Assessment (TA) has been developed since the 1960s as an approach first to explore possible unintended and negative side-effects of technology, to elaborate strategies for dealing with them and to provide policy advice (early warning, see Sec. 3). From the 1980s on the idea of *shaping technology* by early reflection on possible later impacts and consequences of technology was postulated (Bijker & Law, 1994). The adaptation of this social constructivist programme to TA was done within the approach of Constructive Technology Assessment (CTA, cp. Rip et al., 1995). Parallel to this development in the field of TA the Leitbild of sustainable development became a major issue in public debate and scientific research. Against this background it is not surprising that TA took up the challenge to start thinking about shaping technology in accordance with sustainability principles (Weaver et al., 2000). Terms such as "transition management" (Kemp et al., 1998) and "reflexive governance" (Voss et al. 2006) were coined in order to demarcate the need for and approaches to embed sustainability assessments into the consideration of the governance of transformation processes toward sustainable development.

In the meantime TA has been used to assessing sustainability impacts of technology in a manifold of fields (see examples in Sec. 2). Main areas have been the fields of sustainable energy supply technologies, waste disposal, environmental technologies, mobility and transport, and also the exploration of sustainability potentials of new technologies such as nanotechnology and synthetic biology. However, these sustainability assessments differ considerably with respect to the understanding and operationalisation of the Leitbild of sustainable development. This situation makes comparisons between different assessments difficult if not impossible, and technology assessments with incompatible or diverging results may be criticized as being arbitrary. In order to overcome this situation, this chapter aims at introducing and proposing a general framework for sustainability assessments of technology.

There are strong needs to exploit and exhaust opportunities for shaping technology according to sustainability principles (see Sec. 2 and the examples given there) in the framework of sustainability strategies and policies because the production, use and disposal of technology is a highly influential power affecting many sustainability dimensions. Shaping technology with respect to sustainability needs preceding and early sustainability assessments of the technologies under consideration. The main objective of this chapter is to make use of the body of knowledge and experience from the field of technology assessment for sustainability assessments of technology by applying an integrative and transparent understanding of sustainable development. Meeting this objective requires, besides the knowledge about technology assessment (see Sec. 3), also a clear picture of how the Leitbild of sustainable development could be made operable. In this respect the integrative concept of sustainability (Kopfmüller et al., 2001) will be introduced (Sec. 4) and applied to the needs of sustainability assessment of technologies (Sec. 5). In addition, we will describe main methodical challenges of sustainability assessment of technology and give outlines how to meet them (Sec. 6).

2. Shaping technology for sustainable development – the challenge

Generally, a deep-ranging *ambivalence* of the roles of technology in regard to sustainable development can be observed. The relation between technology and sustainable development is usually discussed under contrary aspects: On the one hand, technology is regarded as a *problem* for sustainability and as cause of numerous problems of sustainability, but on the other hand, it is also and directly considered as a *solution* or at least one aspect of the solution of sustainability problems. This ambiguity is the reason for classifying the relation between technology and sustainability as *ambivalent* (Fleischer & Grunwald, 2002).

On the one hand, the use of technologies in modern society has numerous impacts and consequences which conflict with sustainability requirements. This applies for ecological impacts, especially problems with emissions which are harmful for the environment or health and the rapid exploitation of renewable and non-renewable resources. Also in view of social aspects the technological progress causes sustainability problems, such as the consequence of the technical rationalization for the labour market. At the same time the *distribution* of both the possibilities and risks of modern technology often objects the claim for justice of sustainability – for example: industrialized countries are often the beneficiaries of technological innovations, while developing countries are primarily affected by the disadvantages. The "digital divide", describing not only the unequal opportunities to use the Internet in industrialized compared to developing countries but also within industrialized countries is an often quoted example (Grunwald et al., 2006).

On the other hand, there are also many impacts and consequences of technological progress which are *positive* in the sense of sustainability. Well-known examples are the prosperity which has been achieved in many parts of the world and the consequential security of livelihood and quality of life, the successful control of many diseases which were disastrous in former times, food security in many (not all!) parts of the world, and the possibility of global information and communication through the Internet. *Innovative* technologies play a key role in the so-called efficiency strategies of sustainable development (cf. e.g. Weizsäcker et al., 1995). To some extent, modern technologies can already replace conventional technologies and thus contribute to more sustainability (e.g. by fewer emissions and reduced consumption of resources).

This ambivalent relation between technology and sustainability is the starting point for approaches for shaping technology and its societal ways of use (Weaver et al., 2000). These approaches shall be used to realize the positive sustainability effects of innovative technology and minimize or avoid the negative ones in order to contribute through technological progress to a sustainable development in an optimal way. The resulting question is not whether technological progress has positive or negatives effects on sustainability, but how scientific-technological progress and the use of its results has to be designed to achieve positive contributions to a sustainable development. The questions which have to be analysed in this context include:

- How and to which extent can research, development, and use of new technologies contribute to sustainability? How do technology's contributions to sustainability influence other contributions (e.g. of changing lifestyles and a "sustainable consumption")? Within which period of time can the impacts relevant for sustainability be expected?
- Which societal framework conditions can serve as incentive for the development, production, and market integration of innovative technology as a contribution to more sustainability? Which political instruments can support this?
- Which methods can be used to assess whether and to which extent the use of technology can result in more or less sustainability? Which sustainability criteria can be the basis for these assessments and how are they justified? Where are methodological new or further developments necessary, e.g. in life cycle analysis?
- Which standards of comparison, weighing principles, and criteria for consideration can be used in situations of contrary effects and conflicts of aims concerning sustainability?
- How reliable or arguable are sustainability assessments of technology? How should be dealt with the unavoidable uncertainty and ambivalence concerning the knowledge on impacts and assessment problems?

The structure of these questions is very similar to that of the types of tasks of technology assessment (TA) (Grunwald, 2009). In the end it is about *prospectively* understanding and assessing technology impacts relevant for sustainability – preferably already during the *development* of a technology. The principle of considering such knowledge on presumable or probable technology impacts already in the early stages of decision-making and making it thus usable for the design of technology itself or its societal "embedment" is part of the basic concept of TA. Therefore the experience of TA can be used to answer the above-mentioned questions of a prospective sustainability assessment of technology (Fleischer & Grunwald, 2002). In this context the requirements for sustainability assessments and their consideration for decision-making processes are accompanied by considerable conceptual and methodological challenges, even regarding the ambitious concepts of TA. It is no exaggeration to say that the well-known methodological problems of technology assessment concerning both future prospects and evaluation are taken to extremes here.

Several experiences have already been made in applying ideas of TA to shaping technology assessment for sustainable development. There many different opportunities, contexts, situations, stage of development of the respective technologies but also different challenges, obstacles and difficulties. Within this spectrum we can identify two extremes:

a.   The transformation of large infrastructures (such as energy supply, water supply, information and communication, and transport) toward more sustainable structures. The issues at stake in this respect is that usually a *system transformation* will be needed where singular technologies are only parts of the game but where social issues, acceptance, user behaviour, governance, power and control are main elements.

b.   New and emerging science and technologies (NEST) such as nanotechnology, microsystems technologies, converging technologies and synthetic biology are *enabling technologies*: they can lead to a lot of revolutionary developments in many application areas. Therefore, they often show high potentials for supporting strategies of sustainable development – however, most of them are related also with high uncertainties and possible risks.

In the following, two examples are briefly introduced to illustrate the differing challenges with respect to sustainability assessments and the elaboration of strategies: the transformation of the energy system as an example for infrastructures and the case of nanotechnology as an example for NEST cases.

(a) The transformation of the energy infrastructure in conjunction with principles of sustainable development is a considerable challenge. Industrialized countries such as Germany have achieved high standards of energy supply. Energy in the form of electricity, gas, or fuel is reliable and has been more or less available to industrial and private consumers without restriction for decades. Changes in these framework conditions can easily lead to societal controversies. Therefore, transformation processes must always take into account the willingness of customers and users to support these changes and implement behavioral adaptations where required. Sustainability assessments of technology therefore must include the 'social side' of the technologies. The energy supply infrastructure is a *socio-technical system*. It can only fulfill its function if supply and demand are balanced, and if the required changes can be integrated into the existing routines of functioning societal processes, or if new routines can be easily established. Therefore, not only is technical competence necessary for the analysis and design of future (sustainable) energy infrastructures, but so are insights into organizational and societal circumstances such as political-legal framework conditions, economic boundary conditions, individual and social behavior patterns, ethical assessment criteria, and acceptance patterns. In addition, other infrastructures must be co-considered with the energy system: in particular the transport infrastructure (through the development towards e-mobility) and the information infrastructures. The interplay between technical potential, complex social usage patterns, and connected regulation and control processes requires a holistic investigation and an interdisciplinary assessment of the transformation-and-governance strategies aiming at sustainable development. Sustainability assessment of new energy infrastructure elements, therefore, must not be restricted to exploring the supply side and to the provision of technical artifacts (machines, power stations, pipelines etc.). Instead, sustainability assessments must consider also *the societal demand and user side*. Research must bridge disciplinary boundaries between the natural, technical, and social sciences and link technical developments to context conditions of markets, organizational strategies and individual behavior. Multiple interfaces between technical, environmental and social issues have to be taken into account.

(b) The situation in the NEST case is completely different. NEST may change products, systems, and value added chains in many different application fields. There is much more open space for shaping technology because of the early stage of development. The main (research) questions for a sustainability assessment of nanotechnologies are (Fleischer & Grunwald, 2008): Can nanotechnology development and the application of the resulting products, processes and systems be organized in a sustainable – or, at least, *more* sustainable – manner? How can nanotechnologies and their application paths be shaped in a way that they positively contribute to sustainable development? Are there possibilities of shaping nanotechnologies already in early stages of R&D? The application of nanotechnology in products and systems is expected to produce a significant relaxation of the burden on the environment: a saving of material resources, a reduction in the mass of by-products that are a burden on the environment, improved efficiency in transforming energy, a reduction in energy consumption, and the removal of pollutants from the environment (Fleischer & Grunwald, 2008). A number of studies on precisely the issue of the sustainability of nanotechnology have been published in the meantime (e.g., JCP, 2008). However, these developments might have a price. The consequences of the use and release of nanomaterials into the environment are unknown. Although it is not very probable that synthetic nanoparticles in the environment will have long-term effects because of anticipated agglomeration processes, there is no proof available. We do not know about possible long-term effects comparable to the HCFC problem that created the hole in the ozone layer. This situation of high uncertainty and ignorance places a burden of possible risk on future generations while we are exploiting the benefits of nanotechnology today.

Shaping nanotechnologies for sustainable development requires anticipatory sustainability assessments in order to permit distinctions to be made between the more and less sustainable technologies (Fleischer & Grunwald, 2008). Anticipatory assessments of nanotechnology have to cover the entire *life cycle* of the respective technological products or systems. They should include a *temporal integration* and *balancing* of all sustainability effects which might occur during the complete life cycle. For such analyses to contribute to shaping nanotechnologies for more sustainability, they must provide reliable *prospective* life cycle information, such as on health and environmental implications, consumption and production patterns, future developments of lifestyles and markets, and the political and economic framework conditions for the later usage of new technologies. These are only some examples of aspects of the future that need to be known *in advance* in order for reliable life cycle analyses to provide sustainability assessments. A start has already been made toward addressing this challenge for the creation of prospective life cycle assessment. The increasing focus on life cycle assessment as a tool for example in strategy and planning processes, including for long-term issues, and in scenario processes has triggered methodological developments that try combine traditional technology foresight methods with life cycle assessment methods (Schepelmann et al., 2009). Decisive for a comprehensive assessment of nanotechnology or of the corresponding products from a sustainability point of view is that the entire course of the products lifetime is taken into consideration. This extends from the primary storage sites to transportation and the manufacturing processes to the product's use, ending finally with its disposal (Fleischer & Grunwald, 2002). In many areas, however, nanotechnology is still in an early phase of development, so that the data about its life cycle that would be needed for life cycle assessment are far from being available. Empirical research on the persistence, long-term behaviour, and whereabouts of nanoparticles in the environment as well as on their respective consequences would be necessary to enable us to act responsibly in accordance with criteria of sustainable development.

These examples show the high variety of methodological and governance challenges, reaching from societal inertia with respect to the transformation of infrastructures to severe uncertainties in the NEST case, followed by a large gap between far ranging expectations and dystopian fears (Grunwald, 2007). With respect to sustainable development it has to be stated that there is a high diversity of understandings governing the respective sustainability assessments. This diversity endangers the validity of any sustainability assessment because any of them could be regarded arbitrary. Also comparisons between different sustainability assessments will be difficult or impossible because of incommensurable sustainability understandings. A comprehensive framework is still missing – in the following such a framework will be introduced and proposed to be used as an overall sustainability assessment framework (Sec. 4).

3. Technology Assessment and its addresses

Technology Assessment constitutes a scientific and societal response to problems at the interface between technology and society (Grunwald, 2009). It has emerged against the background of various experiences pertaining to the unintended and often undesirable side effects of science, technology, and technicization which, in modern times, can sometimes assume extreme proportions (Bechmann et al., 2007). The types of challenges that have evolved for TA are these: that of integrating at an early stage in decision-making processes any available knowledge on the side effects, that of supporting the evaluation of the value of technologies and their impact, that of elaborating strategies to deal with the knowledge uncertainties that inevitably arise, and that of contributing to the constructive solving of societal conflicts on technology and problems concerning technological legitimization. What characterizes TA is its specific combination of *knowledge production* (concerning the development, consequences, and conditions for implementing technology), the *evaluation* of this knowledge from a societal perspective, and the *recommendations* made to politics and society. The overall aim of TA is to "make a difference" which means to create real influence on the ongoing process of technology development and use. In this sense, sustainability assessment of technologies mustn't be a distant and purely analysing activity but has to be regarded an actor in the field. This situation of TA being an observer and analyst in combination with taking an active role in technology governance results in particular requirements for methodological clarity, quality assurance, and transparency in any respect, in particular regarding the values involved. In addition this situation implies that TA has to be aware of ongoing projects also during the lifetime of TA projects (see Fig. 1).

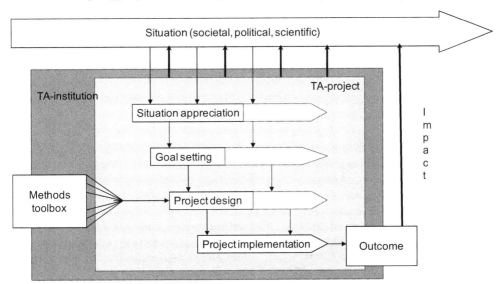

Fig. 1. TA influencing the ongoing societal situation by concrete TA projects continuously keeping track with developments at the societal level (Decker & Ladikas 2004).

Governance of technology has become much more diverse and complex over the past decades. While in earlier times (in the "classical mode" of TA, cp. Grunwald 2009a) a strong role of the state was supposed nowadays much more actors and stakeholders are regarded as being influential on the development and use of new technologies: companies, consumers, engineers, non-governmental organisations (NGO), stakeholders of different kind and citizens. Depending on their roles and occasions to take influence the advice provided by TA could or should look different – in this sense the shift from "steering technology" to a "governance of technology" has had a major influence on TA. Theories of technology development and governance could provide orientation for TA whom to address and what to deliver. Technology assessment with respect to sustainable development will have to consider this variety of actors and stakeholders in being effective and efficient.

The political level remains a major player since governmental technology policy creates obligations for everyone with (partially) high influence on technology. Policy consultation by TA can, for example, take place in the preparatory phase of legislation relevant to technology or even in the very early phases of opinion-forming in the political parties. In the run-up to policy decisions it is possible for TA to carry out enlightenment by reflecting on possible consequences and impacts of technology on society and on the values touched. This positioning of TA research and consultation affects all constellations in which state action influences technology including direct state-run or at least state-dominated technology development, for example in the fields of space travel, military technology, and transportation infrastructure; indirect political influence on technology by means of programs promoting research and technology, for example in materials science, on regenerative sources of energy, or in stem cell research; indirect political control of technology by setting boundary conditions such as environmental and safety standards, laws on privacy or laws stipulating recycling and the role of the state as a *user* of technology, e.g., with regard to the observance of sustainability standards (public procurement). In all of these fields issues relevant to sustainable development play an important role. In some countries this situation already led to the obligation to perform a sustainability assessment of new laws.

TA gives advice to policy-makers in all of these fields and to the involved organisations such as parliaments, governments, and authorities. An example is the *Office of Technology Assessment* at the German *Bundestag* (TAB: http://www.tab-beim-bundestag.de). The purpose of the TAB is to provide contributions to the improvement of the legislature's information basis, in particular, of research- and technology-related processes of parliamentary discussion. Among its responsibilities are, above all, drawing up and carrying out TA projects, and – in order to prepare and to supplement them – observing and analyzing important scientific and technical trends, as well as societal developments associated with them (monitoring). The TAB is strictly oriented on the German *Bundestag's* and its committees' information requirements. The choice of subjects for TA projects as well as their delimitation and specification is the *Bundestag's* responsibility. Decisions on the urgency of problems and the scientific advice desired belong on the political agenda. The subjects of the TAB's studies stem from all fields of technology. The "classical" TA subjects, such as technology and the environment, energy, and bio- and genetic engineering, predominate, involving challenges of sustainable development to an increasing extent over the last years.

The concrete development of technology, however, takes place primarily in the economy at market conditions. The shaping of technology by and in enterprises is operationalized by means of requirement specifications, project plans, and strategic entrepreneurial decisions. These in turn take place on the prescriptive basis of an enterprise's headline goals, general principles, plan goals, and self-understanding but also including assumptions about later consumers and users of the technology and future market conditions. Engineers and engineering scientists have influence on decisions at this level and are confronted in a special way with attributions of responsibility because of their close links with the processes of the development, production, utilization, and disposal of technology. Technology assessment became aware of the importance of this part of technology governance in the 1980s in the course of the social constructivist movement leading to the slogan of "shaping technology" (Bijker et al., 1987; Bijker & Law, 1994). In particular, the approach of Constructive Technology Assessment (CTA, cp. Rip et al., 1995) took up ideas of shaping technology according to the requirements of sustainable development.

The individual preferences of users and consumers of technical systems and products help determine the success of technology developments in two ways: first, by means of their purchasing and consumer behaviour, and second (and less noted), by means of their comments in market research. The influence on technological development resulting from consumer behaviour arises from the concurrence of the actual purchasing behaviour of many individual persons. A well-known problem is, for example, that awareness of a problem with regard to the deficient environmental compatibility of certain forms of behaviour — though definitely present — may not lead to a change in behaviour. Technology assessment aims, in this field, at public enlightenment and information about consequences of consumer's behaviour and at enabling and empowering individuals to behave in a better reflected way, in particular towards more sustainable consumption patterns..

The course of technical development is also decided by public debates, above all by those in the mass media. Public discussion in Germany influenced, for example, political opinion on atomic energy, thus providing much of the basis for the decision in 2002 to phase out atomic energy in that country, and to return to this position after a more positive appraisal of nuclear energy after the Fukushima disaster. Similarly, the public discussion about genetically modified organisms has influenced the regulatory attitude of the European Union and the official acceptance of the precautionary principle. This can also be recognized by the fact that different regulations were established in those countries in which the public debates were very different, such as in the USA. Technology assessment has become an actor also in this field by involving itself in participatory processes playing an increasing role also in political decision-making processes in many countries, in particular in relation with sustainability issues.

4. The integrative sustainability concept

There is considerable need for orientation knowledge on how to fill the Leitbild of sustainable development with substance conclusively as soon as it is expected to guide the transformation of societal systems, e.g. the energy system. To gain practical relevance, some essential criteria have to be fulfilled: (1) a clear *object relation,* i.e. by definition it must be clear what the term applies to and what not, and which are the subjects to which assessments should be ascribed; (2) the *power of differentiation,* i.e. clear and comprehensible differentiations between "sustainable" and "non- or less sustainable" must be possible and concrete ascriptions of these judgements to societal circumstances or developments have to be made possible beyond arbitrariness; (3) the possibility to *operationalize,* i.e. the definition has to be substantial enough to define sustainability indicators, to determine target values for them and to allow for empirical "measurements" of sustainability.

The integrative concept (see Kopfmüller et al., 2001) identified three constitutive elements of sustainable development out of the famous Brundtland Commission's definition (WCED, 1987), taking into account also results of the World Summit at Rio de Janeiro 1992 and from ongoing scientific results and debates. These three elements are (1) the global perspective, (2), the justice postulate and (3) the anthropocentric point of departure.

(1) An essential aspect of sustainable development is, first of all, the global orientation. The assignment given the Commission by the UN General Assembly was that "a global agenda for change" was to be formulated to help define" the aspirational goals for the world

community", and how they could be realized through better cooperation "between countries in different stages of economic and social development" (WCED, 1987). The Commission's report is based on a fresh view of the problems, which interprets the phenomenon of global environmental deterioration and the growing prosperity gap between North and South as interrelated crises of modern industrial society. On the basis of this understanding of the problem, the Commission elaborates the concept of sustainable development as a sound, long-term model for the survival and welfare of global society. The chances for overcoming the global crisis, in their view, depend on the extent to which we succeed in making this concept a model for a "global ethic" (WCED, 1987). On the basis of the general consensus arrived at in the Commission on the prerequisites for global sustainable development, the individual nations are supposed to elaborate on the national level specific targets and strategies for realizing the general objectives, which would be adapted to their own respective current circumstances.

(2) The most important criterion for sustainability is, from the viewpoint of the Brundtland Commission, that of justice. Fundamental for their interpretation of justice is, first of all, the mutual interdependence of intra- and intergenerational justice: a just(er) present is the prerequisite for a just future. In this case, "justice" is primarily defined as distributive justice. The present inequalities of distribution as regards access to natural resources, income, goods and social status is regarded to be the cause of global problems and conflicts, and a juster distribution or rather, a re-distribution of rights, responsibilities, opportunities, and burdens is required. The integrative concept is, following and concretizing this line of thought, based on the postulate that – following Rawls – every human being has a right of access to certain basic goods, which are indispensable preconditions for a self-determined life. Every human being is entitled to these rights and goods, independent of his or her accomplishments, and regardless of circumstances, for which he or she is not to be held accountable. The guarantee of human rights in their entirety is seen as a precondition for, rather than the content of sustainable development. In view of the difficulties of prognosticating the needs of future generations, keeping options open, or, rather, upholding possible choices for future generations is seen as a basic requirement of intergenerational justice. In contrast to other concepts, which regard only the responsibility for future generations as constitutive for sustainability (s. above), inter- and intragenerational justice are here held to be related and, normatively, equal in rank. If one – on the basis of the principle of responsibility for future generations – postulates that access to certain primary goods has to be ensured throughout time, one must also expect that these basic goods are made accessible to all humans today. Otherwise, one would demand – in the interest of future generations – the awarding of rights which are denied to the present generation.

(3) The third element which is characteristic for the Brundtland Report's understanding of sustainability is the anthropocentric orientation. The satisfaction of human needs is, in this concept, the primary goal of sustainable development – today, and in the future. The conservation of nature is not taken as an objective in its own right, but as a prerequisite for lasting societal progress – that is to say, nature is seen as a means to mankind's ends. Humanity is responsible for nature, because humans, as natural beings, are dependent on certain ecosystem services, on the functioning of natural cycles and growth processes. Even when nature is attributed intrinsic value as living space and as a source of experiences, this is done from the viewpoint of, and according to the standards of, human beings. Like most of the other concepts of sustainability, the integrative concept is based on a position of

"enlightened" anthropocentrism which justifies the responsibility for a cautious utilization of nature with mankind's own well-understood self-interest. The responsibility of preserving the diversity of options for human interaction with nature for the future follows out of the postulate that the same rights are to be granted to future generations as the present one enjoys.

These constitutive elements are operationalized further in two steps: first, they were "translated" into the three general goals of sustainable development (1) securing human existence, (2) maintaining society's productive potential (comprising natural, man-made, human, and knowledge capital), and (3) preserving society's options for development and action. In a second step, these goals are concretized by sustainability principles, which apply to various societal areas or to certain aspects in the relationship between society and nature. The concept distinguishes between substantial principles, identifying minimum conditions for sustainable development that ought to be assured for all people living in present and future generations (see table 1a – 1c), and instrumental principles, describing necessary framework conditions for the realization of the substantial minimum conditions (see table 2). On the one hand, these principles – to be further concretized by suitable indicators – unfold the normative aspects of sustainability as goal orientation for future development and as guidelines for action; on the other hand they provide criteria to assess the sustainability performance of particular societal sectors, spatial entities, technologies, policies, etc.

It seems quite important for the understanding of the concept and its consequences for Sustainability Technology Assessment (cp. next Section) to explain the relation between the super-ordinate goals and the principles more in-depth, in particular to give arguments for the determination of the sustainability principles (following Brandl et al. 2002). This will be done in the following. An overview of the substantial principles is given in Figures 1a – 1c. The numbers of the principles mentioned in the following sections refer to the structure of the tables in order to allow for a quick overview.

1.1 Protection of human health	Hazards and unacceptable risks to human health due to anthropogenic environmental burdening must be avoided.
1.2 Ensuring basic needs	Every member of society must be assured a minimum of basic supplies (housing, food, clothing, health care) and protection against fundamental risks to life (sickness, disability).
1.3 Securing an autonomous existence	All members of society must be given the possibility of securing their existence by voluntarily undertaken activities (including education of children and care of the elderly).
1.4 Fair sharing in the use of natural resources	Utilization of natural and environmental resources must be distributed according to the principles of justice and a fair participation of all persons affected.
1.5 Balancing extreme inequalities in income and wealth	Extreme inequalities in the distribution of income and wealth must be reduced.

Table 1. a. Substantial sustainability principles related with the general sustainability objective "Securing human existence" according to the integrative sustainability concept. Source: Kopfmüller et al., 2001 (translated). The left column contains the short title, the right one the principle.

(1) Securing Mankind's Existence

The prime necessity which can be derived from the postulate of justice is, without doubt, that the present generation shouldn't destroy the basis of its own subsistence and that of future generations. A fundamental precondition for this aim is, first of all, that the environmental conditions necessary for human health are upheld. This comprises, in particular, the responsibility of minimizing impacts on the environment which can also impair human health (Principle 1.1). Besides health protection, the satisfaction of basic needs is seen as an indispensable prerequisite for an adequate human existence. These are nourishment, clothing, shelter, basic medical care, access to pure drinking water and to sanitary facilities, as well as safeguards against crucial existential hazards, such as illness, disability, and social crises (Principle 1.2). With respect to sustainable development, the goal can, however, not merely consist in securing "naked existence", but must rather include the best possible preparation for individuals to plan their lives themselves in an active and productive manner. The minimum prerequisite for this empowerment is, that all members of society have the opportunity to secure an adequate and stable existence, including the education of children and provision for old age, by means of an occupation chosen of their own free will (Principle 1.3). While the responsibility of satisfying basic needs is reduced to the material core of the vitally necessary goods, this principle, formulated according to Amarthya Sen, is directed at the presuppositions for a self-determined life. The purpose is to *enable* people to provide themselves with everything they need, instead of their merely being provided through transfer payments or other external assistance. Providing the basis for an independent livelihood presupposes, in its turn, that access to the necessary resources is assured. A necessary condition for this purpose is a just distribution of the opportunities for making use of the globally accessible environmental goods (the earth's atmosphere, the oceans, water, biodiversity, etc.) with the fair participation of all concerned (Principle 1.4). The postulate of ensuring acceptable living conditions and autonomous self-support also implies, finally, that extreme differences in income and wealth be compensated as well as possible (Principle 1.5). This last principle reiterates the previous one with regard to the distribution of income and wealth. This has to be just, at least inasmuch as extreme poverty, which makes active participation in social life impossible, and would lead to social exclusion, has to be precluded.

(2) Maintaining Society's Productive Potential

Future generations should find comparable possibilities of satisfying their needs, which mustn't necessarily be identical to those of the present generation. Regarding the *material* needs, one can derive from this postulate the requirement that the productive capacity of (global) society has to be upheld through time – in a quite general sense – as a generic goal of sustainable development. Every generation disposes over a certain productive potential, which is made up of various factors (natural capital, real capital, human capital, knowledge capital). Sustainable development demands in general, that the stock of capital which exists within a generation be handed down as undiminished as possible to future generations – whereby, however, two fundamentally different alternatives are conceivable (cf. Daly, 1999, p. 110ff.). On the one hand, one could stipulate that the sum of natural and human-made capital be constant in the sense of an economy-wide total; on the other hand, one could require that every single component of itself has to be preserved intact. The former path is sensible if one assumes that natural and human-made capital are interchangeable (*weak* sustainability). The latter path is advisable if one assumes that human-made and natural capital stand in a complementary relationship to one another (*strong* sustainability). The controversy over both of these strategies, that is, over the question, how the heritage which is to be handed down to future generations should be composed, is one of the central problems of the sustainability debate (cf. Ott/Döring 2004). In the integrative concept the substitution of natural capital by human-made capital is held to be admissible to a limited extent, as long as nature's basic functions (the immaterial ones as well) are maintained. With regard to renewable resources, it stipulates that the rate of their use shouldn't exceed the rate of their regeneration, whereby the manner of use is to be taken into consideration along with the intensity of use (Principle 2.1). With reference to non-renewable resources, it assumes that it isn't possible to abstain from their use entirely, but that their consumption has to be compensated. The approach postulates that the range of the known non-renewable resources remains constant through time (Principle 2.2). This principle can only be kept if we either abstain from consuming these resources (sufficiency), if resource productivity is increased (efficiency), if non-renewable resources are replaced by renewable resources (consistency), or if new reserves are tapped. In order to maintain the functions of stabilization and support indispensable for humanity, it requires that the anthropogenic material input should not exceed the absorptive capacity of the environmental media and of the ecosystems (Principle 2.3). To complement these three principles on the use of nature, the integrative concept postulates further that technical hazards with possibly catastrophic effects on human beings and on the environment are to be avoided (Principle 2.4). The formulation of this sort of principle was felt to be necessary, because the risk component is only insufficiently comprehended by the other principles. Setting limits, for example, in fact requires, regarding the maximum pollutant level in environmental media, the weighing of risks; this, however, orients itself, in general, on "trouble-free, normal operation", and leaves the possibility of breakdowns – to a great extent – out of consideration. With regard to the general goal of maintaining society's productive potential, finally, the integrative concept postulates developing real, human and knowledge capital so, that economic efficiency is

2.1 Sustainable use of renewable resources	The rate of utilizing renewable resources is not to exceed the regeneration rate or endanger the ecosystems' capability to perform and function.
2.2 Sustainable use of non-renewable resources	The range of proved non-renewable resources must be maintained.
2.3 Sustainable use of the environment as a sink for waste and emissions	The release of substances is not to exceed the absorption capacity of the environmental media and ecosystems.
2.4 Avoiding unacceptable technical risks	Technical risks with potentially catastrophic impacts on humanity and the environment must be avoided.
2.5 Sustainable development of man-made, human, and knowledge capital	Man-made, human, and knowledge capital must be developed in order to maintain or improve the economy's performance.

Table 1.b. Substantial sustainability principles related with the general sustainability objective "Maintaining society's productive potential" according to the integrative sustainability concept. Source: Kopfmüller et al., 2001 (translated). The left column contains the short title, the right one the principle.

upheld or improved (Principle 2.5). Above all, with regard to real capital, the concept of "development" used here not only includes the possibility of conservation or adaptation, in the sense of building up or restructuring, but, where possible, of reduction as well. The criteria for these decisions follow out of the application of the other principles formulated here.

(3) Keeping Options for Development and Action Open

The precept of not endangering the satisfaction of future generations' needs can, however, not be limited to material necessities but has to include *immaterial* needs as well. For human existence, immaterial aspects such as integration in social and cultural relationships, communication, education, contemplation, aesthetic experiences, leisure, and recreation are just as indispensable as the material bases of subsistence and just as important. Only when these needs have also been satisfied can one speak of a stable and acceptable level of human existence. With regard to the individual human being, this means that the opportunities for personal development have to be secured in the present and for the future. A minimum prerequisite for attaining this goal would be, first of all, the guarantee of equal opportunity in access to education, information, culture, to an occupation, to public office, and to social status (Principle 3.1). Free access to these goods is seen as the basis for equal opportunity for all members of society to develop their own talents and to realize their life plans. As a basic precondition for a self-determined life, equal opportunity is, at the same time, a necessary prerequisite for meeting the demand for autonomously earning one's own living (s. above, Principle 1.3). The second indispensable minimum requirement is the opportunity for participation in societally relevant decision-making processes (Principle 3.2). The basis for this principle is the conviction that a society can only then be considered sustainable – in normative as well as functional respect – when it offers its members the chance for participation in the formation of societal volition. Its purpose is to uphold, broaden, and improve democratic forms of decision-making and conflict management, especially in view of decisions which are of critical importance for the future development and organization of society. In the concept of sustainability, participation is a means as well as an end: with regard to the individual's right to a self-determined life, participation is a goal. Proceeding on the conviction that a process of development in the direction of sustainability can only then be successful, if it is initiated and supported by a broad societal basis, participation is, at the same time, an instrument. With regard to the general goal of not restricting future generations' options for development and action, one would have to raise the further demand that present options also shouldn't be restricted. A minimum requirement for this purpose is that the historical heritage, as well as the diversity of cultural and aesthetic values is preserved (Principle 3.3). This precept includes the protection of nature above and beyond its economic function as a source of raw materials and as a sink for pollutants: nature, resp., certain elements of nature, have to be protected because of their cultural importance as an object of contemplative, intellectual, religious, and aesthetic experiences (Principle 3.4). The minimum requirements listed above primarily refer to the interests of individual members of society, while the aspect of the social system or of society as a whole remained to a great extent left aside. The expectations of individuals with regard to self-actualization and autonomy, however, don't necessarily harmonize with society's demands for integration, stability, and conformity. In the interest of sustainable development, this conflict relationship has to be balanced out. A society which wants to remain lastingly viable has to provide for the integration, socialization, participation, and motivation of its members, and have the capability of appropriate reaction to changed circumstances. A minimum requirement for securing society's cohesion is seen in maintaining its "social resources". This means that tolerance, solidarity, a sense of civility and justice, as well as the capability for the peaceful resolution of conflicts have to be improved (Principle 3.5).

3.1 Equal opportunities	All members of society must have equal chances to access education, occupation, information, and public functions as well as social, political, and economic positions.
3.2 Participation in societal decision-making processes	Every member of society should be given the opportunity to participate in relevant decision-making processes.
3.3 Conservation of cultural heritage and diversity	Human cultural heritage and cultural diversity must be preserved.
3.4 Conservation of the cultural function of nature	Cultivated and natural landscapes or areas of special uniqueness and beauty have to be preserved.
3.5 Conservation of social resources	To ensure societal cohesion, the sense of legal rights and justice, tolerance, solidarity, and perception of common welfare as well as the possibility of non-violent conflict settlement must be enhanced.

Table 1.c. Substantial sustainability principles related with the general sustainability objective "Preserving development and action options" according to the integrative sustainability concept. Source: Kopfmüller et al., 2001 (translated). The left column contains the short title, the right one the principle.

The basic orientation of the fifteen substantial principles is influenced by the three general sustainability objectives assigned to them. The general objective 'Securing human existence' focuses on the individual as being the prime beneficiary. The general objective 'Maintaining society's productive potential' refers to the indispensable prerequisites of various societal activities and is by no means limited to the material prerequisites for the conventional production of goods and services in the private and public sector. According to the general objective 'Preserving development and action options' the current generation is, if it can, required to establish and preserve the prerequisites for the freedom of decision by future generations.

The fifteen substantial principles *collectively* represent minimum requirements and may be complemented by additional requirements, provided the original principles are not violated. The substantial principles may be fulfilled to different degrees. If, however, two of them are in conflict, they will have to be weighed up. In the general model, the instrumental principles are indispensable and equal the substantial principles.

Conflicts of goals between principles can exist on different levels. First of all, it cannot be excluded that the formulated working hypothesis of a simultaneous satisfiability of all principles will be falsified. Undiminished population growth, for instance, could lead to such a falsification, if satisfaction of basic needs of the world population would not be possible without breaking e.g. the natural resource-related principles. Other conflict potentials can arise when the guiding principles are translated into concrete responsibilities of action for societal actors. In such conflicts, each principle can be valid only within the limits set by the others. Additionally, the concept includes a weighing principle by distinguishing between a core scope for each principle which always has to be fulfilled and may not be weighed against other principles, and a rather peripheral scope where weighing is possible. Regarding for instance the principle "Ensuring satisfaction of basic needs", the core scope would be the pure survival of everyone, whereas the peripheral scope would have to be defined, to a certain extent according to particular regional contexts.

Internalization of environmental and social external costs	Environmental and social external costs arising in an economic process must be considered within this process.
Appropriate discounting	Discounting may not discriminate against future or present generations.
Limiting public indebtedness	In order to avoid restricting the state's future action and design scope, in principle current public consumption expenditures must be funded by current income.
Fair global economic framework	The global economic framework conditions should be designed so that economic actors of all states have the fair chance to participate in economic processes.
Enhancing international co-operation	The various actors (governments, enterprises and non-governmental organizations) must co-operate in the spirit of global partnership in order to create the political, legal, and factual prerequisites for implementing sustainable development.
Society's ability to respond	Society's ability to respond to problems in the natural and societal spheres must be enhanced by suitable institutional innovations.
Society's ability to reflect	Institutional conditions must be developed, which allow reflection on societal action options beyond isolated problems and individual aspects.
Ability to steer	Society's ability to steer towards a sustainable development must be increased.
Self-organization	The self-organization potential of societal actors has to be enhanced.
Balancing power	Opinion-building, negotiation, and decision processes must be designed so that the possibilities of societal actors to express themselves and exert influence are justly distributed and the processes are transparent.

Table 2. Instrumental principles according to the integrative sustainability concept. Source: Kopfmüller et al., 2001 (translated)

The conflict potential included in the sustainability principles shows that even an integrative concept is not harmonistic. Rather, the integrative nature of sustainability increases the number of relevant conflicts. This approach is able to uncover those – otherwise hidden – conflicts in defining and implementing sustainable development. Thus, conflicts are by no means to be avoided but rather are at the heart of any activities to make sustainability work (Grunwald, 2005). Rational conflict management and deliberation are, therefore, of great importance.

Sustainable development remains a political and normative notion also in the scientific attempts of clarification and operationalization. Therefore, it will not be possible to provide a kind of "algorithm" for sustainability assessments allowing for calculating an objective "one best solution" of sustainability challenges. What can be done, however, is to clarify the framework for assessments and societal decision-making to support transparent, well-informed, and normatively-orientated societal processes of deliberation on sustainability.

5. Sustainability principles to be applied in technology assessment

The integrative sustainability concept has not been specifically developed as an instrument for technology assessment but refers to the development of society as a whole in the global perspective. However, technology is always just one component of societal relations and developments; many other and sometimes more relevant aspects – like patterns of production and consumption, lifestyles, and cultural conventionalities, but also national and global political framework conditions – have to be considered to understand and assess societal developments. If the integrative sustainability concept is used as normative framework for technology assessment, it has to be kept in mind that technology can only make (positive as well as negative) *contributions* to a sustainable development (Weaver et al., 2000). Moreover, these contributions always have to be seen against the background of other societal developments. Energy technologies as such are neither sustainable nor unsustainable but can only make more or less large contributions to sustainability – or cause problems.

First of all it has to be determined which principles of sustainability are relevant for technology assessment. The following principles can prima facie be considered relevant in the energy context: Protection of human health, securing the satisfaction of basic needs, sustainable use of renewable resources, sustainable use of non-renewable resources, sustainable use of the environment as a sink, avoidance of unacceptable technical risks, participation in societal decision-making processes, equal opportunities, internalization of external social and environmental costs and society's reflexivity. Characteristic aspects of the relation of these principles to technology will be described in the following, including the wording of the principle (for a more detailed explanation see Kopfmüller et al., 2001).

Protection of human health

Dangers and intolerable risks for human health due to anthropogenically-caused environmental impacts have to be avoided. Production, use, and disposal of technology often have impacts which might negatively affect human health both in the short or long term. On the one hand this includes accident hazards in industrial production (work accidents), but also in everyday use of technology (the large number of people injured or killed are a sustainability problem of motorized road traffic). On the other hand, there are also "creeping" technology impacts which can cause harmful medium- or long-term effects by emissions into environmental media. The history of the use of asbestos and its devastating health effects are a particular dramatic example from the working environment (Gee & Greenberg, 2002). However, there are also – at least in industrialized countries – the big successes in combating diseases or the prolongation of the human life expectancy due to medical progress or sanitary supply and disposal technologies. Also food preservation technologies and the resulting improvement of nutrition are positive effects.

Securing the satisfaction of basic needs

A minimum of basic services (accommodation, nutrition, clothing, health) and the protection against central risks of life (illness, disability) have to be secured for all members of society. Technology plays an outstanding role in securing the satisfaction of basic human needs through the economic system; energy supply is also essential for this. This applies directly for the production, distribution, and operation of goods to satisfy the needs (e.g. technical infrastructure for the supply of water, energy, mobility, and information, waste and sewage disposal, building a house, household appliances). However, this is on the one hand opposed by numerous negative impacts resulting from this way of need satisfaction common in industrialized countries (which then show up against the background of other sustainability principles). On the other hand, it has to be kept in mind that a large part of the world population is still cut off from this basic satisfaction of needs secured by means of technology. For example, approx. 2 billion people do not have access to a regular energy supply. Some 1.2 billion people worldwide have no adequate drinking water supply. 2.4 billion people are not connected to a safe and hygienic wastewater disposal.

Sustainable use of renewable resources

The usage rate of renewable resources must neither exceed their replenishment rate nor endanger the efficiency and reliability of the respective ecosystem. Renewable natural resources are e.g. renewable energies (wind, water, biomass, geothermal energy, solar energy), ground water, biomaterials for industrial use (e.g. wood for building houses) and wildlife or fish stock. In the historical development of the concept of sustainability the principle on renewable resources has played a major role in the context of forestry and fishery. It contains two statements. On the one hand, it is essential that resources are extracted in a gentle way to protect the inventory. Human usage shall not consume more than can be replenished. On the other hand, it has to be ensured that the respective ecosystems are not overstrained, e.g. by emissions or serious imbalances. Here technology plays an important role in using the extracted resources as efficient as possible (e.g. energetic use of biomass) and minimizing problematic emissions.

Sustainable use of non-renewable resources

The reserves of proven non-renewable resources have to be preserved over time. The consumption of non-renewable resources like fossil energy carriers or certain materials calls for a particularly close link to technology and technological progress. The consumption of non-renewable resources may only be called sustainable if the temporal supply of the resource does not decline in the future. This is only possible if technological progress allows for such a significant increase in efficiency (von Weizsäcker et al., 1995) of the consumption in the future that the reduction of the reserves imminent in the consumption does not have negative effects on the temporal supply of the remaining resources. So a *minimum speed* of technological progress is supposed. The principle of reserves directly ties in with efficiency strategies of sustainability; it can be really seen as a *commitment* to increase efficiency by technological progress and respective societal concepts of use for the consumption of non-renewable resources. One alternative, which also depends on the crucial contributions of technological concepts, would be *substituting* non-renewable resources in production and use of technology with renewable ones (e.g. the reorganization of the energy supply for transport from mineral oil to electricity from regenerative sources). Regarding the material resources, the ideal of recycling management includes the idea of recycling the used materials to the largest extent and in the best quality possible; thus the available resources would hardly decline in amount and quality. However, this ideal reaches its limits, since recycling normally includes high energy consumption and material degeneration.

Sustainable use of the environment as a sink

The release of substances must not exceed the absorption capacity of the environmental media and ecosystems. Extraction of natural resources, processing of materials, energy consumption, transports, production processes, manifold forms of use of technology, operation of technical plants, and disposal processes produce an enormous amount of material emissions which are then released into the environmental media water (ground water, surface water, and oceans), air, and soil. These processes often cause serious regional problems, especially concerning the quality of air, ecosystems, biodiversity, and freshwater. However, environmental measures taken in industrialized countries led to considerable progress in this respect over the last decades. Unfortunately, this does neither apply for most developing and newly industrializing countries nor for global effects like degradation of soil used for agriculture, accumulation of persistent pollutants in polar seas, or the release of greenhouse gases. Technology plays a major role in all strategies for solving these problems. On the one hand, as an "end-of-pipe" technology it can reduce the emissions at the end of technical processes, e.g. in form of carbon capture and storage (CSS). On the other hand, and this is the innovative approach, technical processes can be designed in a way that unwanted emissions do not occur at all. This requirement usually results in a significant need for research and development which even extends to basic research.

Avoidance of unacceptable technical risks

Technical risks with potentially disastrous impacts for human beings and the environment have to be avoided. This principle is necessary since the way to handle disastrous technical risks is insufficiently described in the three "ecological management principles". These management principles refer to "failure-free normal operation" and disregard possible incidents and accidents as well as unintended "spontaneous side effects". They are rather intended for long-term and "creeping" processes like the gradual depletion of natural resources or the gradual "poisoning" of environmental compartments. The risk principle refers to three different categories of technical risks: (1) risks with comparatively high occurrence probability where the extent of the potential damage is locally or regionally limited, (2) risks with a low probability of occurrence but a high risk potential for human beings and the environment, (3) risks that are fraught with high uncertainty since neither the possibility of occurrence nor the extent of the damage can currently be sufficiently and adequately estimated. This principle is closely linked to the precautionary principle (by von Schomberg, 2005). It could be applied to the problems discussed in the context of a severe nuclear reactor accident (worst-case scenario such as recently happened at Fukushima,

Japan), for securing the long-term safety of a final repository for highly radioactive waste, or possible risks of the release of genetically modified organisms.

Conservation of nature's cultural functions

Cultural and natural landscapes or parts of landscapes of particular characteristic and beauty have to be conserved. A concept of sustainability only geared towards the significance of resource economics of nature would ignore additional aspects of a "life-enriching significance" of nature. The normative postulate to guarantee similar possibilities of need satisfaction to future generations like the ones we enjoy today can therefore not only be restricted to the direct use of nature as supplier of raw materials and sink for harmful substances but has to include nature as subject of sensual, contemplative, spiritual, religious, and aesthetic experience. Within the energy context one has to be reminded of the final repository for radioactive waste at Yucca Mountain in the United States, where problems occurred due to the spiritual meaning of the region to the indigenous population. Also the changing landscapes due to wind farms are a problem in some regions; this is discussed not only in connection with tourism but also regarding the aesthetic values of landscapes.

Participation in societal decision-making processes

All members of society must have the opportunity to participate in societally relevant decision-making processes. Regarding technology, this principle has a substantial and a procedural aspect (see in general Joss & Belucci, 2002). On the one hand (substantially) it affects the design of technologies which (might) be used for participation. Here the principle calls for exploiting these potentials of participation as far as possible. On the other hand (procedurally) the principle aims at the conservation, extension, and improvement of democratic forms of decision-making and conflict resolution, especially regarding those decisions which are of key importance for the future development and shaping of the (global) society; the aspect of designing future energy systems is definitely part of this. Future energy supply, far-reaching ethical questions of biomedical sciences with probably significant cultural impacts, questions of risk acceptance and acceptability in case of genetically modified food are examples for technological developments with a considerable sustainability relevance which should be – according to this principle – dealt with participative methods.

Equal opportunities

All members of society must have equal opportunities regarding access to education, information, occupation, office, as well as social, political, and economic positions. The free access to these goods is seen as prerequisite for all members of society to have the same opportunities to realize their own talents and plans for life. This principle primarily relates to questions of societal organization where technology only plays a minor role. However, the availability of energy is often a crucial precondition for being able to participate in societal processes at all, e.g. for having access to information and communication technologies which need energy or mobility which is also impossible without energy. The fact that approx. 2 billion people in the world do not have access to a regular energy supply underlines the circumstance that this also considerably restricts their possibilities of participation. Furthermore, the access to information has to be mentioned as another central challenge. The call for equal opportunities regarding the access to societally relevant information includes expectations of the technical infrastructure (e.g. technical enabling of Internet access by being connected to a suited communication network), but also the requirements to the competences of the user to be able to handle this technical infrastructure and use it accordingly. Concerning the effects of Internet use on democracy, there were high hopes that the Internet could revive or renew democracy. The facilitated access to information and the new options for communication were expected to lead to an "empowerment", i.e. they would turn citizens who are only little informed and disenchanted with politics into active and well-informed citizens (cf. Grunwald et al., 2006). These visions turned out to be illusions, which was – apart from other reasons (Grunwald et al., 2006) – also due to the fact that the realization of positive hopes concerning democracy required easy Internet access for everyone not just "in theory" but also "in reality", a precondition which has by far not been fulfilled. This does not only apply for the disparity between Internet use in industrialized and developing countries, but is also a problem of developed countries where e.g. elderly people or people from disadvantaged groups often cannot deal with this technology with its high innovation rate (both aspects together are often referred to as "digital divide").

Internalization of external social and environmental costs

Prices have to reflect the external environmental and social costs arising through the economic process. One reason for the neglect of essential ecological and social aspects in the economic process, for suboptimal allocation of resources, and the resulting sustainability deficits lies in the fact that only one part of the costs arising from the production and consumption process is considered for pricing. The so-called "external effects" or "external costs" refer to the effects of production and consumption activities which are not borne by the causer but by third parties and are a priori not subject to a regulation via market or pricing mechanisms. External effects lead to distortions of prices and the structure of goods and therefore also to discrepancies between commercial and societal costs and/or between micro and macro rationalities of the market process which contributes to the development of sustainability deficits. The call for cheap energy or cheap materials often result in not taking the "real" costs as a basis since negative impacts for the environment, health, and future generations are not implied. Exempting fuel from the tax liability is one example for the lacking internalization of such external costs.

Appropriate discounting

Discounting shall neither discriminate against today's nor future generations. Discounting procedures are used to make effects in the form of economically relevant quantities occurring at different times comparable and assessable for current decision-making processes. In doing so, cost and benefit items which result from investments and other activities in the course of the given period are discounted to their current or cash value. So the question is how much a subsequent loss or benefit is worth compared to today's losses or benefits. Answers to this question are crucial for long-term developments. Relevant examples include the determination of strategies to deal with the climate change, with highly radioactive wastes which have to be controlled for millenniums and burden future generations with costs and possible risks. The question of the appropriate discount rate cannot be decided ethically or scientifically but only politically – albeit in accordance with scientific information and ethical orientation.

Society's reflexivity

Institutional arrangements have to be developed, which make a reflection of options of societal action possible, which extend beyond the limits of particular problem areas and individual aspects of problems. Technology is future-oriented. It is developed in relation to aims and functions and shall realize certain technological functions and performance characteristics which are not yet achievable today. The intended aims of technology and the subsequent real impacts of its use are not always identical (Grunwald, 2009). If the technological development is finished and the respective technology is implemented or used, the relation to the future changes. Then technology no longer has only anticipated but also real impacts. The aims pursued with technology may be achieved only partially, others might be exceeded. Technologies which were expected to be a promising and profitable solution turn out to be commercially unsuccessful while functionalities which were rather developed as a "by-product" are the big winner. Impacts of the use of certain technologies, either feared or hoped for, do not occur at all or not in the expected extent; there are side effects which nobody anticipated. When it comes to technology design for sustainability, possible non-intended impacts have to be taken into account early enough. Given the far-reaching societal impacts of technology, the instrumental sustainability principle on reflexivity calls for (a) strengthening the awareness for impacts, conducting impact research and impact reflection, and sensitizing societal subsystems (especially policy, economy, and science) for this, (b) establishing a comprehensive and multi-perspective view on the impacts instead of just focussing on specific fields of impact, and (c) providing enough resources and time for impact reflection in societal opinion-making and decision-making processes. Technology assessment as a process accompanying technology development (Grunwald, 2009) can help to achieve this.

Sustainability principles cannot be directly transferred into guidelines for technology design or even performance characteristics for technology. They do not refer to technological requirements but to aspects of society's economic behaviour where technology is just one aspect among others. If the consequences for technology are in the focus, the *context* has to be taken into consideration: Which are the problems relevant for sustainability in the respective field, which technological and which societal conditions apply, how are they connected, and how does the whole (and often quite complex) structure relate to the approach of the whole system of sustainability principles. So the sustainability principles have by no means a prescriptive character for technology design. A number of steps of transfer and mediation have to be done on the way from normative orientation to concrete technology design. This task cannot be in the sole responsibility of the people involved in technology development. In particular cases societal dialogues are necessary and, where appropriate, even political decisions. Exactly this situation, where the system of sustainability principles provides orientation without determining technology in detail, supports the theory that the sustainability postulate is suitable as Leitbild for technology design. However, there are – and this will be discussed in the following – sometimes considerable conceptual and methodological challenges.

6. Methodological challenges and the learning cycle

Technology assessment as contribution to societal technology design has three different aspects (Grunwald, 2009): Provision of knowledge by research on technology and impacts of technology, analysis of their normative implications (assessment), and societal communication in the light of upcoming processes of opinion- and decision-making (advice). The best available knowledge provided by different scientific disciplines has to be considered for the design of energy technology regarding sustainability aspects. Moreover, a broad, ethically enlightened societal dialogue on the goals of design, on visions of a future society, on desirability, acceptability, and reasonability of future developments, on risks, and on the distribution of opportunities and risks is necessary. Concerning TA as sustainability assessment, this intensifies the challenges already known from other fields of TA: The spectrum which has to be considered is enlarged regarding its content, the timeframe, and by the increase of criteria relevant for the assessment. In performing such assessments and especially in providing the required knowledge, some typical and serious methodical challenges have to be dealt with (Grunwald, 2010):

- *Embeddedness of technology*: Technology as such is neither sustainable nor unsustainable. The contributions of technology to sustainability are co-determined by technology inherent and social or institutional elements; the ways technologies are used and embedded into society exert a strong influence (Weaver et al., 2000). Assessments have to take into consideration technological and societal processes, structures, values, customs, etc. that might be affected by the way of technology embeddedness into society.
- *The life cycle approach and the prediction problem*: Anticipatory sustainability assessments have to cover the entire *life cycle* of e.g. technologies or products, including resource mining, transport and treatment processes, the societal use processes, the impacts for both the natural environment and society, and finally the disposal phase. In sustainability assessments "prospective life cycle analyses" are needed, taking properly into account future consumer and production patterns (Brown et al., 2000), lifestyles, markets, or political and economic framework conditions – in order to provide suitable life-cycle analyses based knowledge for sustainability decisions (Schepelmann et al., 2009).
- *The completeness issue and the incompleteness problem*: Due to the broadness and complexity of sustainable development and its normative character, it is impossible for philosophical, economic, and pragmatic reasons to gain a definite complete picture in investigating the sustainability performance of certain subjects. Relevant sustainability aspects might be simply overseen or wrongly deemed of low relevance. Thus, decisions have to be made on the relevance or irrelevance of particular issues and on the boundaries of the systems considered.
- *The integration issue and the incommensurability problem*: Broadness and complexity of sustainable development also lead to another methodological issue: the heterogeneity of the various dimensions and indicators means that no common measure of sustainability can be applied to all subjects of investigation as a whole. Methodically, it is not satisfactory to measure for instance CO_2 emissions, numbers of people affected by long-term unemployment, development co-operation activities, or the engagement in civil society organizations according to the same unique scale. The various sustainability dimensions and indicators cannot be integrated into one single measure like a "sustainability index" without incurring severe methodological problems. As the sustainability imperative is integrative by nature, any attempt to integrate data or assess results from heterogeneous aspects of "overall" sustainability demonstrates normative dimensions which strongly limit the use of common decision-analysis tools. Rather, discursive tools must be used, and political decisions made.

All these issues and challenges result in the necessity to deal with high uncertainties in approaching sustainable development (Grunwald 2004; 2008) and require an increasingly reflexive type of governance (Voss et al., 2006). The uncertainty and incompleteness of knowledge and the provisionality of evaluations make a complete implementation of sustainability in the sense of a detailed planning in the direction of sustainability impossible. The respective current state of the art in knowledge production plays an unfathomable role by the elaboration of sustainability strategies and assessments of current developments. It becomes obvious with all clarity that a policy of sustainability has to be carried out under conditions of uncertain knowledge and of provisional assessments. A policy of sustainability is therefore confronted with the limits of the availability of knowledge. It is *ex ante* not stringently decidable, whether and to what extent a political measure, a technological innovation, or a new institutional arrangement will, in actual application, "really" contribute to sustainability. Every sustainability policy has to face this situation and become – in a certain sense – "experimental" (Grunwald 2004). This situation also applies to the development of technology with regard to the requirements of sustainable development and has the result that the classical procedure, in which scientific knowledge is simply implemented or "applied", is not practicable in sustainability policy. The traditional approach, the production of knowledge through research and the application of this knowledge by politics, misses the problematic point just mentioned. Instead, knowledge and action intertwine: the road leads not only from knowledge to action (orienting approach), but also from action to knowledge (experimental approach). Precisely this situation, in combination with the impossibility of a detailed planning in the direction of sustainability, suggests understanding the relationship between science and politics in regard to implementation strategies as a process of learning (see Fig. 2).

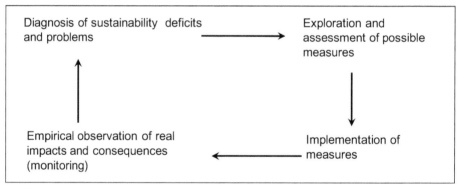

Fig. 2. Development towards more sustainability as a feedback loop involving, measures, empirical measurement and reflection. The integrative concept of sustainable development can be used for orientating diagnoses and assessments with respect to their normative substance.

For a purposive sustainability policy, it is decisive that, in this experimental situation, the inadequacy and incompleteness of knowledge doesn't paralyze or hinder action, but that, in the interpretation and implementation of practical measures, a maximum range of opportunities for learning in these "experiments" is seized. Contributions of scientific research to these learning processes consist, to put it allegorically and remain in the metaphor of the experiment, in first ensuring as good a preparation of the experiment as possible (by analyses of the situation, by causal analyses, by modelling and simulation of proposed measures, etc.), in supervising the careful execution of the experiment, and then in observing the results of the process, in comparing them with the goals pursued, and, if necessary, investigating the reasons for deviations.

7. Conclusions

The integrative approach to understand sustainability per definition without hastily reducing it to merely ecological aspects has proven the richness of the spectrum of aspects of sustainability. Of course criteria of resource economics and ecology are of special importance. But also questions of participation and equal opportunities, the way to deal with technical risks and aesthetic values of landscapes, the shaping of reflexive societal decision processes and the modelling of economic framework conditions as well as aspects of human health play crucial roles. Compared with this result it has to be noted that the sustainability debate on technologies in industrial countries is often narrowed to ecological issues, at the utmost supplemented by aspects of economic development or social peace. In contrast, it has to be pointed out: the sustainability of technologies has to be measured against a much larger spectrum of principles, criteria, and indicators than often assumed.

However, this spectrum aggravates the well-known problems of *prospective* sustainability assessment (Sec. 5). Especially with regard to unavoidable conflicts of objective between the different criteria of sustainability and the incommensurability of many criteria the need for a methodologically secured approach of sustainability assessment is obvious. Classical instruments like life-cycle assessment or simulations are required, but by no means sufficient. On the one hand, they have to be developed further to meet the range of sustainability criteria. Approaches like consequential LCA or social LCA veer towards this, but are of course just starting off. On the other hand, qualitative procedures of deliberation for "soft" criteria of sustainability and for the consideration of conflicts of objectives are necessary. The concept introduced in this paper does not solve these methodological problems; but nevertheless it provides a well-founded conceptual framework for the further development of these methods of assessment on a transparent basis.

Sustainability assessments of technological options, technology impacts, or innovation potentials therefore include considerable and ineliminable uncertainties. Especially assessments are made *under uncertainty*. This applies on the one hand for assessment criteria which are themselves subject to change over time (thinking, e.g., of the emergence of ecological awareness in the 1970s and its consequences for assessment processes). Assessments are also made relatively to the *state of knowledge* and are therefore dependent on uncertainty, incompleteness, and preliminarity of this knowledge. Hence the problem of knowledge (see above) has immediate effects on the assessment question. The assessment of asbestos for example changed abruptly when carcinogenic effects were discovered; in the same way the assessment of chlorofluorocarbons changed after the discovery of the mechanism which caused the ozone hole. The consequence is that *design* supported by TA as sustainability assessment of technology cannot be understood as a planning towards a determined goal but as a permanent *process* considering societal dialogues and learning processes.

The *knowledge question* (Grunwald, 2004) and the *assessment question* make a *final* sustainability assessment of technology impossible. The question whether sustainability will be achieved can only be answered in connection with the use and the embedment of technology into society. Technology can contribute more or less to a sustainable economic behaviour but cannot decide on sustainability alone. Sustainability assessments and the resulting decisions depend on the context, are preliminary, and open to further development as a result of societal learning processes: steps in a co-evolution of society and technology on a "sustainable" way into the future (Grunwald, 2008).

It is about using the manifold possibilities to understand technology design under sustainability aspects as a *permanent* learning process: as a societal process where design objectives and options for realization are being discussed, which is influenced by scientific knowledge and ethical orientations, where the concept of a "sustainable" technology is developed step by step. In this way, technology assessment is a medium of learning, where technology development, the development of the respective societal framework conditions, and the use of technology are critically accompanied as well as analyzed and assessed under sustainability aspects.

8. References

Bechmann, G., Decker, M., Fiedeler, U., Krings, B. (2007). TA in a complex world. International Journal of Foresight and Innovation Policy 4, p. 4-21

Bijker, W. E., Hughes, T. P., Pinch, T. J. (eds.) (1987). The Social Construction of Technological Systems.. Cambridge (Mass.)

Bijker, W., Law J. (eds.) (1994). Shaping Technology/Building Society. Cambridge (Mass.)

Brandl, V., Jörissen, J., Kopfmüller, J., Paetau, M. (2002). Das integrative Konzept nachhaltiger Entwicklung. In: A. Grunwald et al.. Forschungswerkstatt Nachhaltigkeit. Berlin, pp. 73-102

Brown, N., Rappert, B., Webster, A. (eds.) (2000). Contested Futures. A sociology of prospective techno-science. Burlington/Ashgate, 2000

Brown-Weiss, E. (1989). In Fairness to Future Generations. International Law, Common Patrimony and Intergenerational Equity. New York

Daly, H. (1987): The Economic Growth Debate: What Some Economists Have Learned But Many Have Not. In: Journal of Environmental Economics and Management, Vol. 14, S. 323–336

Daly, H. (1999): Wirtschaft jenseits von Wachstum. Die Volkswirtschaftslehre nachhaltiger Entwicklung. Salzburg

Daly, H. E. (1994): Operationalizing Sustainable Development by Investing in Natural Capital. In: Jansson, A.; Hammer, M.; Folke, C.; Costanza, R. (Eds.): Investing in Natural Capital. The Ecological Economics Approach to Sustainability. Washington D.C., S. 22–37

Decker, M., Ladikas, M. (eds.) (2004). Bridges between Science, Society and Policy. Technology Assessment – Methods and Impacts. Berlin

Fleischer, T., Grunwald, A. (2002). Technikgestaltung für mehr Nachhaltigkeit – Anforderungen an die Technikfolgenabschätzung. In: Grunwald, A. (ed.): Technikgestaltung für eine nachhaltige Entwicklung. Von der Konzeption zur Umsetzung. Berlin, p. 95–146

Fleischer, T., Grunwald, A. (2008). Making nanotechnology developments sustainable. A role for technology assessment? *Journal of Cleaner Production* 16(2008), p. 889-898

Gee, D., Greenberg, M. (2002). Asbestos: from 'magic' to malevolent mineral. In: Harremoes, P., Gee, D., MacGarvin, M., Stirling, A., Keys, J., Wynne, B., Guedes Vaz, S. (Hg.): The Precautionary Principle in the 20th century. Late Lessons from Early Warnings. London, p. 49-63

Grunwald, A. (2004). Strategic knowledge for sustainable development: the need for reflexivity and learning at the interface between science and society. *International Journal of Foresight and Innovation Policy* 1(2004)1/2, pp. 150-167

Grunwald, A. (2005). Conflicts and Conflict-solving as Chances to Make the Concept of Sustainable Development Work. In: P. A. Wilderer, E. D. Schroeder, H. Kopp (eds.): Global Sustainability. The Impact of Local Cultures. A New Perspective for Science and Engineering, Economics and Politics. Weinheim: Wiley-VCH, p. 107-122

Grunwald, A. (2007). Converging Technologies: visions, increased contingencies of the conditio humana, and search for orientation. Futures39, p. 380-392

Grunwald, A. (2008). Working Towards Sustainable Development in the Face of Uncertainty and Incomplete Knowledge. *Journal of Environmental Policy & Planning* Volume 9, Issue 3, pp. 245-262

Grunwald, A. (2009). Technology Assessment: Concepts and Methods. In: A. Meijers (ed.): *Philosophy of Technology and Engineering Sciences*. Volume 9. Amsterdam: North Holland 2009, p. 1103-1146

Grunwald, A. (2010). Technikfolgenabschätzung. Eine Einführung. Berlin, second edition

Grunwald, A., Banse, G., Coenen, C., Hennen, L. (2006). Netzöffentlichkeit und digitale Demokratie. Tendenzen politischer Kommunikation im Internet. Berlin

Grunwald, A.; Kopfmüller, J. (2006). Nachhaltigkeit. Eine Einführung. Frankfurt a. Main

JCP - Journal of Cleaner Production (2008). Sustainable Nanotechnology Development. Special Issue. Journal of Cleaner Production 16

Joss, S.; Bellucci, S. (2002). Participatory Technology Assessment in Europe: Introducing the EUROPTA Research Project. In: Joss, S.; Bellucci, S. (Eds.): Participatory Technology Assessment. European Perspectives. Center for the Study of Technology, Westminster, pp. 3-14

Kemp, R, Shot J & Hoogma, R (1998). Regime Shifts to Sustainability Through Processes of Niche Formation: The Approach of Strategic Niche Management. In: *Technology Analysis & Strategic Management*, vol.10, pp.175-195.

Kopfmüller, J. (ed.) (2006). Ein Konzept auf dem Prüfstand. Das integrative Nachhaltigkeitskonzept in der Forschungspraxis. Berlin

Kopfmüller, J., Brandl, V., Jörissen, J., Paetau, M., Banse, G., Coenen, R., Grunwald, A. (2001). Nachhaltige Entwicklung integrativ betrachtet. Konstitutive Elemente, Regeln, Indikatoren. Berlin

Paschen, H., Petermann, Th. (1992). Technikfolgenabschätzung – ein strategisches Rahmenkonzept für die Analyse und Bewertung von Technikfolgen. In: Petermann 1992, S. 19–42

RHM (2011): How sustainable is Santiago de Chile? *Synthesis report of the Risk Habitat Megacity research initiative. Leipzig*

Rip, A., Misa, T., Schot, J. (eds.) (1995). Managing Technology in Society. London

Schepelmann, P., M. Ritthoff, H. Jeswani, A. Azapagic, K. Suomalainen (2009). Options for deepening and broadening LCA. CALCAS - Co-ordination Action for innovation in Life-Cycle Analysis for Sustainability. Brüssel et al.

Sen, A. (1987): On Ethics and Economics. Oxford

Sen, A. (1993): Capability and Well-Being. In: Nussbaum, M.; Sen, A. (eds.): The Quality of Life. Oxford, S. 30–53

von Schomberg, R. (2002). The objective of Sustainable Development: Are we coming closer? EU Foresight Working Papers Series 1, Brussels

von Schomberg, R. (2005). The Precautionary Principle and its normative challenges. In: E. Fisher, J. Jones, R. von Schomberg (Hg.): The precautionary principle and public policy decision making, Cheltenham, UK, S. 161-175

von Weizsäcker, E.U., Lovins, A.B., Lovins, L.H. (1995). Faktor vier. Doppelter Wohlstand – halbierter Naturverbrauch. München

Voss, J.-P., Bauknecht, D. & Kemp, R. (2006, eds.). Reflexive Governance for Sustainable Development, Cheltenham

WCED - World Commission on Environment and Development (1987). *Our common future,* Oxford

Weaver, P., Jansen, L., van Grootveld, G., van Spiegel, E. & Vergragt, P. (2000). Sustainable Technology Development, Sheffield

4

Sustainable Engineering and Eco Design

Chaouki Ghenai
Ocean and Mechanical Engineering Department, Florida Atlantic University
USA

1. Introduction

The material consumption in the United States of America now exceeds ten tones per person per year. The average level of global consumption is about eight times smaller than this but is growing twice as fast. The materials and the energy needed to make and shape them are drawn from natural resources: ore bodies, mineral deposits, and fossil hydrocarbons. The demand of natural resources throughout the 18th, 19th and early 20th century appeared infinitesimal (Ashby et al., 2007, Alonso et al., 2007, Chapman and Roberts, 1983, and Wolfe, 1984). There is also a link between the population growth and resource depletion (Ashby et al., 2007, Alonso et al., 2007, Chapman and Roberts, 1983, and Wolfe, 1984). The global resource depletion scales with the population and with per-capita consumption (Ashby et al., 2007, and Alonso et al., 2007). Per capita consumption is growing more quickly.

The first concern is the resource consumption. Speaking globally, we consume roughly 10 billion tones of engineering materials per year. We currently consume about 9 billion tones per year of hydrocarbon fuels (oil and coal). For metals, it appears that the consumption of steel is the number one (~ 0.8 billion tones per year) followed by aluminum (10 millions tones per year). The consumption of steel exceeds, by a factor of ten all other metals combined. Polymers come next: today the combined consumption of commodity polymers polyethylene (PE), polyvinyl chloride (PVC), polypropylene (PP) and polyethylene-terephthalate, (PET) begins to approach that of steel (see figure 1). The really big ones, though, are the materials of the construction industry. Steel is one of these, but the consumption of wood for construction exceeds that of steel even when measured in tones per year, and since it is a factor of 10 lighter, if measured in m3/year, wood totally eclipses steel. Bigger still is the consumption of concrete, which exceeds that of all other materials combined as shown in Figure 1. The other big ones are asphalt (roads) and glass.

The second concern is the energy and carbon release to atmosphere caused by the production of these materials as shown in Figure 2. This is calculated by multiplying the annual production by the embodied energy of the material (MJ/Kg – energy consumed to make 1 Kg of material). During the primary production of some materials such as metals, polymers, composites, and foams the embodied energy is more than 100 MJ/Kg and the CO_2 foot print exceeds 10 Kg of CO_2 per Kg of materials.

New tools are needed to analyze these problems (high resource consumption, energy use and CO_2 emissions) best material based on the design requiment but also to reduce the environmental impacts.

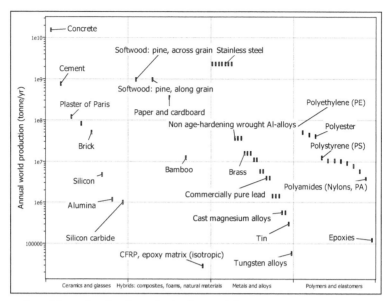

Fig. 1. Annual world production for principal materials

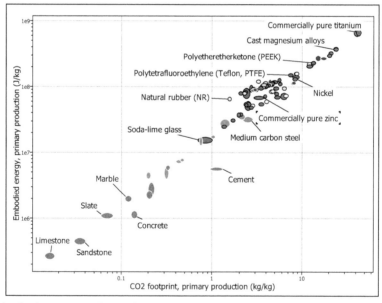

Fig. 2. Embodied Energy and CO2 footprint - primary production of principle materials

To select an eco friendly and sustainable material, one need to examine first the materials life cycle and consider how to apply life cycle analysis (Ashby et al., 2007). The materials life cycle is sketched in Figure 3. Ore and feedstock are mined and processed to yield materials. These materials are manufactured into products that are used and at the end of life, discarded, recycled or (less commonly) refurbished and reused. Energy and materials are consumed in each phase (material, manufacturing, use, transportation and disposal) of life, generating waste heat and solid, liquid, and gaseous emissions (Ashby et al., 2007). The results of the eco audit or life cycle analyis is shown in Figure 4. The results of the life cycle analysis will reveal the dominant phase that is consuming more energy or producing high CO2 emission. The next step is to separate the contributions of the phases of life because subsequent action depends on which is the dominant one. If it is that a material production, then choosing a material with low embodied energy is the way forward. But if it is the use phase, then choosing a material to make use less energy-intensive is the right approach – even if it has a higher embodied energy.

This chapter introduces the methods and tools that will guide in the design analysis of the role of materials and processes selection in terms of embodied energy, carbon foot print, recycle fraction, toxicity and sustainability criteria. A particular skills need to be used by engineer or designer to guide design decisions that minimize or eliminate adverse eco impacts. Methods and tools that will guide in the design analysis of the role of materials and processes selection in terms of embodied energy, carbon foot print, recycle fraction, toxicity and sustainability criteria need to be used during the design process. Topics covered in this chapter will include: resource consumption and its drivers, materials of engineering, material property charts, the material life cycle, eco data, eco-informed material selection, and eco audits or life cycle analysis. The Cambridge Engineering Selecor software (Granta Design Limited, 2009) is used in this study for better understanding of these issues, create

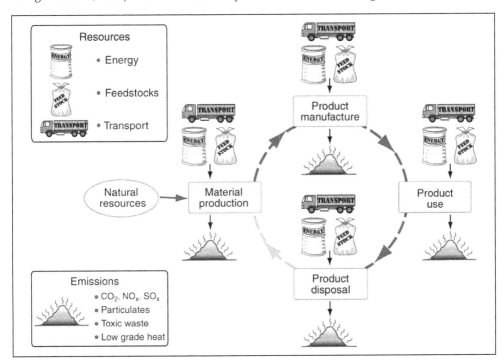

Fig. 3. Material Life Cycle (Ashby et al., 2007)

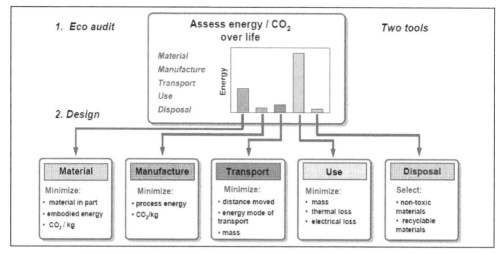

Fig. 4. Eco Audit and Eco Design (Ashby et al., 2007)

material charts, perform materials and processes selection, and eco audit or life cycle analysis allowing alternative design choices to meet the engineering requirements and reduce the environmental burden. The results of two case studies (material selection of desalination plant heat exchanger and life cycle analysis of patio heater) will be presented in this chapter book.

2. Material and process families and eco data

The common material properties are: general properties (cost and density), mechanical properties (strength, stiffness, toughness), thermal properties (conductivity, diffusivity, expansion, heat capacity), electrical properties (electrical conductivity, dielectric constant), optical (refraction, absorption), magnetic, and chemical properties (corrosion resistance). Materials properties determine the suitability of a material based on design requirements. A successful product, one that performs well, is good value for money and gives pleasure to the user uses the best materials for the job, and fully exploits its potential and characteristics. Materials selection is not about choosing a material, but a profile of properties that best meets the needs of the design (Ashby et al., 2007, and Alonso et al., 2007). Material and process are interdependent and grouped into families; each family has a characteristic profile (family likeness) which is useful to know when selecting which family to use for a design. In general, there are six families for materials (Ashby, 2005): metals (steels, cast irons, alloys...), ceramics (alumina, silicon carbides), polymers (polyethylene, polypropylene, polyethylene-terephthalate), glasses (soda glass, borosilicate glass), elastomers (isoprene, neoprene, butyl rubber, natural rubber) and hybrids (composites, foams) as shown in Figure 5.

Processes are also classified based on the design requirements (material, shape, dimensions, precision, and the number of parts to be made). The process families (Ashby, 2005) are: shaping (casting, molding, deformation, machining, heat treatment), joining (fastening, welding, adhesives) and surface treatments (polishing, painting) as shown in Figure 6.

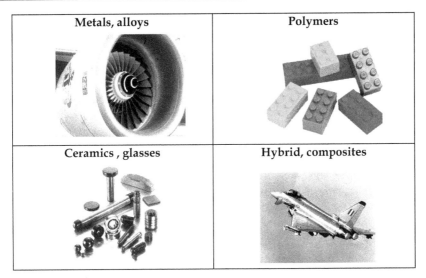

Fig. 5. Materials Families

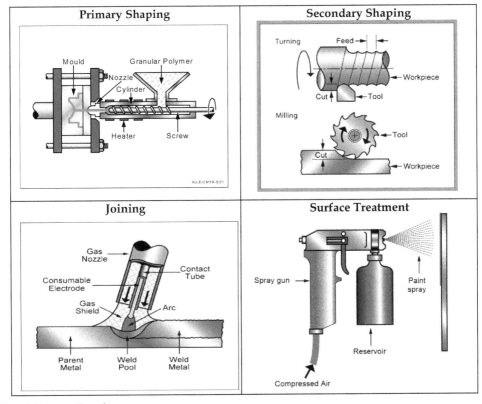

Fig. 6. Process Families

The material and process selection based on some design requirements rely on the materials mechanical, thermal, electrical and chemical properties. Rational selection of materials to meet environmental objectives starts by identifying the phase of product-life that causes greatest concern: production, manufacture, use or disposal. Dealing with all of these requires data for the obvious eco-attributes such as energy, CO2 (Chapman, 1983) and other emissions, toxicity, ability to be recycled and the like (see table 1). Thus if material production is the phase of concern, selection is based on minimizing production energy or the associated emissions (CO2 production for example). But if it is the use-phase that is of concern, selection is based instead on light weight, excellence as a thermal insulator, or as an electrical conductor (while meeting other constraints on stiffness, strength, cost etc). Additional information such eco data (embodied energy and CO2 foot print as shown in Figure 2 and table 1) is needed for sustainable engineering and eco design.

Geo-Economic Data for Principal Component

Annual world production	21e6– 23e6	tonne/year
Reserves	2e10 – 2.2e10	tonne
Typical exploited ore grade	30 – 34	%

Material Production – Energy and Emissions

Production energy	190 – 210	MJ/Kg
CO2	12-13	kg/kg
NOX	72-79	g/kg
SOX	120- 140	g/kg

Indicators for Principal Component

Eco indicator	740 – 820	mmillions points/kg

End of life

Recycle	True	
Down cycle	True	
Biodegrade	False	
Incinerate	False	
Landfill	True	
Recycle as fraction	34 – 38	%

Bio Data

Toxicity rating	Non toxic	
Approve for skin and food contact	True	

Sustainability

Sustainable material	No	

Table 1. Eco Data – Wrought Aluminium Pure

3. Life cycle analysis and selection strategies

The material life cycle is shown in Figure 3. Ore and feedstock, drawn from the earth's resources, are processed to give materials. These materials are manufactured into products that are used, and, at the end of their lives, discarded, a fraction perhaps entering a recycling loop, the rest committed to incineration or land-fill (Gabi, 2008, Graedel, 1998, and Kickel, 2009) Energy and materials are consumed at each point in this cycle (phases), with an associated penalty of CO_2, SO_x, NO_x and other emissions, heat, and gaseous, liquid and solid waste. These are assessed by the technique of life-cycle analysis (LCA) (Ashby, 2007).

3.1 The steps for life cycle analysis are:

1. Define the goal and scope of the assessment: Why the assessment needs to be done? What is the subject and which part of its life are assessed?
2. Compile an inventory of relevant inputs and outputs: What resources are consumed? (bill of materials) What are the emissions generated?
3. Evaluate the potential impacts associated with those inputs and outputs
4. Interpretation of the results of the inventory analysis and impact assessment phases in relation of the objectives of the study: What the result means? What needs to be done about them?

The study examine the energy and material flows in raw material acquisition; processing and manufacturing; distribution and storage (transport, refrigeration...); use; maintenance and repair; and recycling options.

3.2 The strategy for guiding design

The first step is to develop a tool that is approximate but retains sufficient discrimination to differentiate between alternative choices. A spectrum of levels of analysis exist, ranging from a simple eco-screening against a list of banned or undesirable materials and processes to a full life cycle analysis, with overheads of time and cost.

The second step is to select a single measure of eco-stress. On one point there is some international agreement: the Kyoto Protocol of 1997 committed the developed nations that signed it to progressively reduce carbon emissions, meaning CO_2 (Kyoto Protocol, 1999). At the national level the focus is more on reducing energy consumption, but since the energy consumption and CO_2 production are closely related, they are nearly equivalent. Thus there is certain logic in basing design decisions on energy consumption or CO_2 generation; they carry more conviction than the use of a more obscure indicator. We shall follow this route, using energy as our measure.

The third step is to separate the contributions of the phases (material, manufacturing, use, transportation and disposal) of life because subsequent action depends on which is the dominant one with respect of energy consumption and CO_2 emissions (see Figure 4).

For selection to minimize eco-impact we must first ask: which phase of the life cycle of the product under consideration makes the largest impact on the environment? The answer guides material selection. To carry out an eco-audit or life cycle analysis we need the bill of material, shaping or manufacturing process, transportation used of the parts of the final product, the duty cycle during the use of the product, and also the eco data for the energy and CO_2 footprints of materials and manufacturing process.

4. Results

Two case studies of sustainable engineering and eco design are presented in this chapter book. The first case study deals with material selection for the condenser used in desalination plant (sustainable material). The question is what is the best material that can be used for the condenser based on some constraints and design objectives? The second case study is about the life cycle analysis of patio heater. The question is what the dominant phase of the life cycle of this product that is consuming more energy and producing more CO_2 emissions?

4.1 Case study 1: Material selection for desalination plant heat exchanger

Desalination of seawater is one of the most promising techniques used to overcome water shortage problems (Nafey et al., 2004). The desalination techniques include thermal desalination processes (Multi Stage Flash - MSF, Multi Effect Distillation – MED) and membrane desalination processes (reverse osmosis – RO and Electro-Dialysis Reverse - EDR). Multi Stage Flash (MSF) is one of he most commonly distillation process used for large-scale desalination of seawater (Hassan, 2003). In the MSF process, the seawater enters the evaporation chamber resulting in flash boiling of a fraction of the seawater. The vapour produced by flashing is then conveyed to the heat recovery section where it is condensed. Heat exchanger (evaporator and condensers) tubes represent the largest item in an MSF plant and not surprisingly more than 70% of the corrosion failures in desalination plants are attributed to heat exchange tubes. Heat exchangers tubes handle two fluids of completely different properties (seawater and vapors). It is one of the severest environments from the point of view of corrosion (Anees et al., 1993, and Aness et al., 1992). This study focuses only on the desalination plant condensers. The condenser is a sea water-cooled shell and tube heat exchanger installed in the exhaust steam from the evaporator in thermal desalination plant. The condenser is a heat exchanger that converts the steam received from the evaporator to liquid using the sea water as the cooling fluid. The key properties of the desalination plant surface condenser are: (1) heat transfer properties (thermal conductivity, convective heat transfer coefficients for steam and sea water, and fouling coefficients); (2) the erosion resistance (to steam for the external surface of the tube, and to raw sea waters which may contain sand and show turbulences for the internal surface of the tube); (3) corrosion resistance (to raw sea waters, steam and condensate). The heat transfer performance of the condenser is linked to the material selection – thermal conductivity, thickness, and the erosion/corrosion resistance of the tubing materials.

A condenser with high tubing thermal conductivity, thin wall tubing, and tubing surface that do not corrode in the heat exchanger environment and remains relatively cleans during the desalination process will provide excellent heat transfer performance. The principal objective of this study is to select the best materials for the condenser tubing (sustainable material) that will provide excellent thermal heat transfer performance, low cost and low embodied energy (sustainable energy) and CO_2 foot print (sustainable environment).

The condenser shown in Figure 7 takes heat from the steam and passes it to the sea cooling water. The steam enters the shell at temperature T_V, changes its phase from gas to liquid during the heat transfer with the sea cooling water and exit the heat exchanger as condensate at temperature T_C. The sea water cooling fluid enters the condenser tubes at

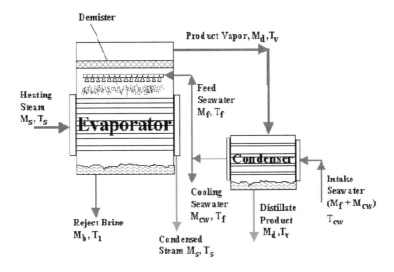

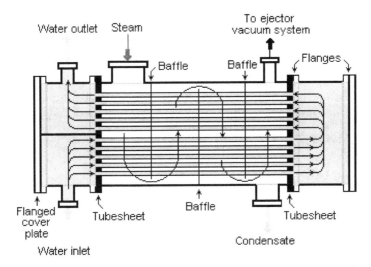

Fig. 7. Desalination process and heat exchanger (condenser)

temperature T_{CW} and exit at high temperature T_{HW}. A key element in all heat exchangers is the tube wall or membrane which separates the sea water and the steam. It is required to transmit heat and there is frequently a pressure difference across it Δp (pressure difference between the sea water and the steam pressures). The question is what are the best materials for making these condensers? What are the best condenser materials that can provide high thermal conductivity but at the same time can sustain this pressure difference? What is the performance index that can be use for heat exchanger or condensers? The heat transfer from the steam to the sea water through the membrane or the thin wall involves convective transfer from steam to outside surface of the condenser tubes, conduction through the tube wall, and convection again to transfer the heat to sea water. The heat flux q into the tube wall by convection (W/m^2) is described by the heat transfer equation $q = h_1\,\Delta T_1$, where h1 is the heat transfer coefficient for the steam and $\Delta T1$ is the temperature drop across the surface from the steam into the outside tube wall. Conduction is described by the conduction equation; $q = (\lambda\,\Delta T_{12})/e$, where λ is the thermal conductivity of the wall (thickness e) and $\Delta T12$ is the temperature difference across the tube wall. The heat flux q out from the tube wall by convection is described by the heat transfer equation $q = h_2\,\Delta T_2$, where h_2 is the heat transfer coefficient for sea water and ΔT_2 is the temperature drop from the inside surface of the tube to the sea water. The heat flux is also given by: $q = U\,(T_V\text{-}\,T_{CW})$, where U is the overall heat transfer coefficient and T_V is the steam temperature entering the shell and T_{CW} is the temperature of sea water entering the tube. The overall heat transfer coefficient is given by:

$$U = \cfrac{1}{\cfrac{1}{h_1} + \cfrac{e}{\lambda} + \cfrac{1}{h_2}} \tag{1}$$

The total heat flow is given by:

$$Q = q\,A = \left(\cfrac{1}{\cfrac{1}{h_1} + \cfrac{e}{\lambda} + \cfrac{1}{h_2}}\right) A\ \left(T_V - T_{CW}\right) \tag{2}$$

One of the constraints of the heat exchanger is that the wall thickness must be sufficient to support the pressure difference Δp. This requires that the stress in the wall remain below the elastic limit (yield strength): $\sigma = \dfrac{\Delta p\,r}{e}\langle\,\sigma_{el}$.

Where r is the pipe radius and e is the pipe thickness.

The heat flux is given by:

$$\frac{Q}{A} = q = \left(\cfrac{1}{\cfrac{1}{h_1} + \cfrac{\Delta p\,r}{\lambda\,\sigma_{el}} + \cfrac{1}{h_2}}\right)\ \left(T_V - T_{CW}\right) \tag{3}$$

The heat flow per unit area of tube wall, Q/A or q is maximized by maximizing the performance index M given by $M = \lambda\,\sigma_{el}$. The maximum value of M is obtained by minimizing the tube wall thickness or maximizing both the thermal conductivity and the yield strength.

Selecting materials for desalination plant heat exchanger involves seeking the best match between design requirements and the properties of the materials that may be used to make the heat exchanger. The strategy for selecting the material for desalination plant heat exchangers is:

a. Translate design requirements: develop a list of requirements the material must meet, expressed as function (what does the system do), objectives (what is to be maximized or minimized), constraints (what nonnegotiable conditions must met) and free variables (what parameters of the problem is the designer free to change). The main function of the condenser is to exchange heat between the steam and seat water (heat exchanger) and to convert the steam to distilled water. The objectives are to maximize heat flow per unit area, minimize the cost and eco friendly materials (minimize the energy and the CO_2 footprint). The constraints for the condenser are: (a) operating temperature up to 150∘C; (b) support pressure difference Δp, (c) excellent resistance to sea water, (d) very high resistance of the material to pitting and crevice corrosion, and (e) excellent resistance of the material stress corrosion cracking. The free choices for the condenser design are the choice of material.

b. Screening: After developing the list of requirements the material must meet, the next step is to eliminate the materials that can not do the job because one or more their attributes lies outside the limits set by the constraints. The limit and tree stages of the Cambridge selector software (Granta Design Limited, 2009) are used in this study as selection tools for the screening process. The limit stage applies numeric and discrete constraint. Required lower or upper limits for material properties are entered into the limit stage property boxes. If a constraint is entered in the minimum box, only materials with values greater than the constraint are retained. If it is entered in the Maximum box, only materials with smaller values are retained. The graph option can be used to create bar charts and bubble charts. A box selection isolates a chosen part of a chart. Any material bar or bubble lying in, or overlapping the box is selected and all others are rejected. The line selection divides a bubble chart into two regions. The user is free to choose the slope of the line, and to select the side on which materials are to be chosen. This allows selection of materials with given values of combinations of material properties such as E/ρ, where E is Young's modulus and ρ is density. The tree stage allows the search to be limited to either: a subset of materials (metals, hybrids, polymers, and ceramics) or materials that can be processes in chosen ways (manufacturing process).

c. Ranking: Find the screening materials that do the job best. Rank the materials that survive the screening using the criteria of excellence or the objectives and make the final materials choice.

Figure 8 shows the results of the screening process for the performance index M. Only 16 materials passed the test based on the design requirements (operating temperature > 150 C, resistance to sea water, resistance to pitting and crevice corrosion, and excellent resistance to stress corrosion cracking). Based on the objectives (maximize heat flux, minimize the cost, the embodied energy and CO_2 foot print) set during the design process, it is clear that the best material that can be used for the desalination plant condenser is the stainless steel duplex UNS S32550, wrought. It has the maximum value for performance index M (high thermal conductivity and thin tube wall), and lowest cost (14-14 $/Kg) as shown in Figure 8. In addition to that this material has the lowest embodied energy and CO_2 foot print as shown in Figure 9. The characteristics of the selected material for the desalination plant condenser are summarized in Table 2.

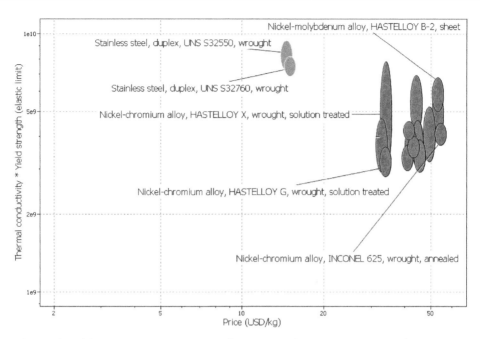

Fig. 8. Results of the Screening Process – Performance Index M versus Material Cost

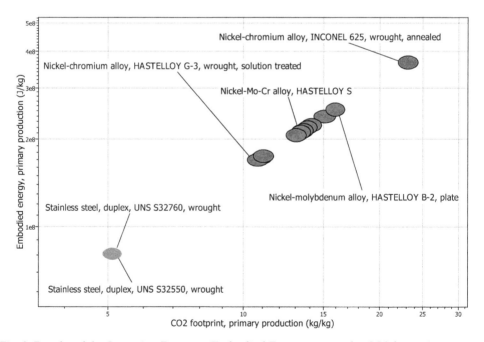

Fig. 9. Results of the Screening Process – Embodied Energy versus the CO2 foot print

Material	Performance Design $M = \lambda\,\sigma_{el}$	Price $/Kg	Pitting and Crevice Corrosion	Stress Corrosion Cracking	Embodied Energy (J/Kg)	Maximum Service Temp. (C)
1. Stainless Steel, Duplex UNS S32550, wrought	7.3e9 – 9.4e9	13.2 – 15.2	Very High	Excellent	7.7e7 – 8.5e7	335 – 365
2. Stainless Steel, Duplex UNS S32760, wrought	6.9e9 – 8.1e9	14.2 – 15.7	Very High	Excellent	7.7e7 – 8.5e7	335 – 365

Table 2. Selected Materials for the desalination plant condenser

4.2 Case Study 2: Life cycle analysis of patio heater

An eco audit is a fast initial assessment. It identifies the phases of life – material, manufacture, transport, and use – that carry the highest demand for energy or create the greatest burden of emissions. It points the finger, so to speak, identifying where the greatest gains might be made. Often, one phase of life is, in eco terms, overwhelmingly dominant, accounting for 60% or more of the energy and carbon totals. This difference is so large that the imprecision in the data and the ambiguities in the modeling, are not an issue; the dominance remains even when the most extreme values are used. It then makes sense to focus first on this dominant phase, since it is here that the potential innovative material choice to reduce energy and carbon are greatest.

An energy and CO2 eco audits were performed for the patio heater shown in Figure 10. It is manufactured in Southeast Asia and shipped 8,000 Km to the United States, where it is sold and used. It weighs 24 kg, of which 17 kg is rolled stainless steel, 6 kg is rolled carbon steel, 0.6 kg is cast brass and 0.4 kg is unidentified injection-molded plastic (See Materials - Tables 3 and 4). During the use, it delivers 14 kW of heat ("enough to keep 8 people warm") consuming 0.9 kg of propane gas (LPG) per hour, releasing 0.059 kg of CO2 /MJ.

The heater is used for 3 hours per day for 30 days per year, over 5 years, at which time the owner tires of it and takes it to the recycling depot (only 6 miles / 10 km away, so neglect the transport CO2) where the stainless steel, carbon steel and brass are sent for recycling (See end of life - Tables 3 and 4). These data are used to construct a bar-chart for energy and CO2 emission over the life of the patio heater.

The table (See Figure 10) lists the energy and carbon footprints of the materials and manufacturing processes for the patio heater. The bar chart plots the totals for each phase. For the sea transport over 8000km, the energy consumed is 30.7 MJ and the CO2 released is 2.18 kg of carbon dioxide, so small as to be invisible on the bar chart. The results show that 97.9% of the energy consumed and 98.1 % of the CO2 emitted are during the use phase. The energy consumed and CO2 emitted for the material phase are respectively 5.9% and 5.2%. The results also show that 4.1% of the energy can be recovered and 3.7 % reduction of CO2 emission can be obtained by recycling the parts of the patio heater. A detailed breakdown of the energy and CO2 foot print for individual life phases (material, manufacture, transport, use, and end of life) are shown respectively in Tables 3 and 4.

Energy and CO2 Footprint Summary:

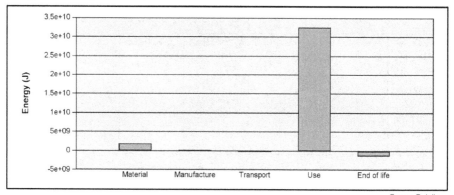

Energy Details...

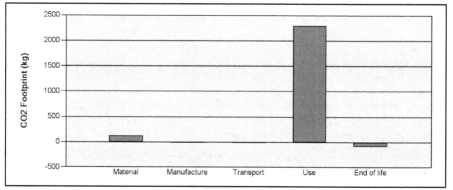

CO2 Details...

Phase	Energy (J)	Energy (%)	CO2 (kg)	CO2 (%)
Material	1.94e+09	5.9	123	5.2
Manufacture	8.41e+07	0.3	6.7	0.3
Transport	3.07e+07	0.1	2.18	0.1
Use	3.24e+10	97.9	2.3e+03	98.1
End of life	-1.36e+09	-4.1	-86.2	-3.7
Total	3.31e+10	100	2.35e+03	100

Fig. 10. Life Cycle Analysis of Patio Heater: Energy and CO2 Footprint Analysis

Material	**Breakdown by component**						
	Component	Material	Recycle content	Material Embodied Energy * (J/kg)	Total Mass (kg)	Energy (J)	%
	Component 1	Stainless steel, duplex, ASTM CD-4MCu, cast	Virgin (0%)	8.1e+07	17	1.4e+09	71.1
	Component 2	Stainless steel, martensitic, ASTM CA-15, cast, tempered at 315°C	Virgin (0%)	8.1e+07	6	4.9e+08	25.1
	Compoent 3	Brass, CuZn10Pb3Sn2, sand-cast	Virgin (0%)	7e+07	0.6	4.2e+07	2.2
	Component 4	PP (65-70% barium sulfate)	Virgin (0%)	7.7e+07	0.4	3.1e+07	1.6
	Total				24	1.9e+09	100

Manufacture	**Breakdown by component**					
	Component	Process	Processing Energy (J/kg)	Total Mass (kg)	Energy (J)	%
	Component 1	Forging, rolling	2.9e+06	17	5e+07	59.6
	Component 2	Forging, rolling	4.1e+06	6	2.5e+07	29.5
	Component 3	Casting	2.7e+06	0.6	1.6e+06	1.9
	Component 4	Polymer molding	1.9e+07	0.4	7.6e+06	9.0
	Total			24	8.4e+07	100

Transport

Breakdown by transport stage   Total product mass = 24 kg

Stage Name	Transport Type	Transport Energy (J/kg.m)	Distance (m)	Energy (J)	%
	Sea freight	0.16	8e+06	3.1e+07	100.0
Total			8e+06	3.1e+07	100

Breakdown by components   Total transport distance = 8e+06 m

Component	Total Mass (kg)	Energy (J)	%
Component 1	17	2.2e+07	70.8
Component 2	6	7.7e+06	25.0
Compoent 3	0.6	7.7e+05	2.5
Component 4	0.4	5.1e+05	1.7
Total	24	3.1e+07	100

Use

Relative contribution of static and mobile modes

Mode	Energy (J)	%
Static	3.2e+10	100.0
Mobile	0	
Total	3.2e+10	100

Static Mode

Energy Input and Output Type	Fossil fuel to thermal, vented system
Product Efficiency	0.7
Use Location	United States
Energy Equivalence, source (J/J)	1
Power Rating (kW)	14
Usage (hours per day)	3
Usage (days per year)	30
Product Life (years)	5
Total Life Usage (hours)	4.5e+02

End of Life	**Relative contributions of end of life options**						
	Component	End of Life Route	Collection Energy (J/kg)	Potential End of Life 'Saving' (J/kg)	Total Mass (kg)	Total EoL Energy (J)	%
	Component 1	Recycle	7e+05	-5.8e+07	17	-9.8e+08	71.7
	Component 2	Recycle	7e+05	-5.8e+07	6	-3.5e+08	25.3
	Compoent 3	Recycle	7e+05	-5e+07	0.6	-2.9e+07	2.5
	Component 4	Landfill	2e+05	0	0.4	8e+04	0.5
	Total				24	-1.4e+09	100

Table 3. Detailed Breakdown of individual life phases: *Energy Analysis* - Patio Heater

Material	**Breakdown by component** 	Component	Material	Recycle content	Material CO2 Footprint * (kg/kg)	Total Mass (kg)	CO2 Footprint (kg)	%
---	---	---	---	---	---	---
Component 1	Stainless steel, duplex, ASTM CD-4MCu, cast	Virgin (0%)	5.1	17	87	70.5
Component 2	Stainless steel, martensitic, ASTM CA-15, cast, tempered at 315°C	Virgin (0%)	5.1	6	31	24.9
Compoent 3	Brass, CuZn10Pt3Sn2 sand-cast	Virgin (0%)	6.2	0.6	3.7	3.0
Component 4	PP (65-70% barium sulfate)	Virgin (0%)	4.7	0.4	1.9	1.5
Total				24	1.2e+02	100	
Manufacture	**Breakdown by component** 	Component	Process	Processing CO2 (kg/kg)	Total Mass (kg)	CO2 Footprint (kg)	%
---	---	---	---	---	---
Component 1	Forging, rolling	0.24	17	4	59.8
Component 2	Forging, rolling	0.33	6	2	29.6
Compoent 3	Casting	0.16	0.6	0.097	1.4
Component 4	Polymer molding	1.5	0.4	0.61	9.1
Total			24	6.7	100	
Transport	**Breakdown by transport stage** Total product mass = 24 kg 	Stage Name	Transport Type	Transport Energy (J/kg.m)	CO2 Footprint, source (kg/J)	Distance (m)	CO2 Footprint (kg)	%
---	---	---	---	---	---	---
	Sea freight	0.16	7.1e-08	8e+06	2.2	100.0
Total				8e+06	2.2	100 | **Breakdown by components** Total transport distance = 8e+06 m | Component	Total Mass (kg)	CO2 Footprint (kg)	%
---	---	---	---
Component 1	17	1.5	70.8
Component 2	6	0.55	25.0
Compoent 3	0.6	0.055	2.5
Component 4	0.4	0.036	1.7
Total	24	2.2	100	
Use	**Relative contribution of static and mobile modes** 	Mode	CO2 Footprint (kg)	%
---	---	---
Static	2.3e+03	100.0
Mobile	0	
Total	2.3e+03	100	 **Static Mode** 	Energy Input and Output Type	Fossil fuel to thermal, vented system
Product Efficiency	0.7
Use Location	United States
CO2 Footprint, source (kg/J)	7.1e-08
Power Rating (kW)	14
Usage (hours per day)	3
Usage (days per year)	30
Product Life (years)	5
Total Life Usage (hours)	4.5e+02	
End of Life	**Relative contributions of end of life options** 	Component	End of Life Route	Collection CO2 (kg/kg)	Potential End of Life 'Saving' (kg/kg)	Total Mass (kg)	Total EoL CO2 (kg)	%
---	---	---	---	---	---	---
Component 1	Recycle	0.042	-3.7	17	-62	71.7
Component 2	Recycle	0.042	-3.7	6	-22	25.3
Component 3	Recycle	0.042	-4.6	0.6	-2.7	2.5
Component 4	Landfill	0.012	0	0.4	0.0048	0.5
Total				24	-86	100	

Table 4. Detailed Breakdown of individual life phases: *CO2 Foot Print* - Patio Heater

5. Conclusion

The methods and tools presented in this book chapter, will guide in the design analysis of the role of materials and processes selection in terms of embodied energy, carbon foot print, recycle fraction, toxicity and sustainability criteria. A particular skills need to be used during the design process not only to satisfy the design requirements but also to minimize or eliminate adverse eco impacts (sustainable design). Two case studies of sustainable engineering and eco design are presented in this chapter book. The first case study deals with material selection for the condenser used in desalination plant and the second case study is about the life cycle analysis of patio heater. The results of the selection process for the heat exchanger (condenser) of a desalination plant show that the best material that can be used for the condenser is the stainless steel, duplex UNS S32255O, wrought. This material has (1) the highest design performance M (high heat flux), (2) the lowest cost (13 – 15 $/Kg), (3) a very good resistance to pitting and crevice resistance, (4) an excellent resistance to stress corrosion cracking (no breaks at high strengths or > 75% of yield strength in various environments), (5) excellent material resistance to sea water (no degradation in material performance expected after a long exposure to sea water), and (6) a good pitting resistance equivalent number (PREN = 40). In addition the embodied energy (energy required to make 1 Kg of the material) and the CO_2 foot print (mass of CO_2 released during the production of 1 Kg of the material) are very low compared to the other materials. The second case study was about the life cycle analysis of the patio heater. The life cycle analysis strategy has two part: (1) an eco audit for a quick and approximate assessment of the distribution of energy demand and carbon emission over the patio heater's life; and (2) material selection to minimize the energy and carbon over the full life, balancing the influence of the choice over each phase of the life (selection strategies and eco informed material selection –suatianble design). The results of the life cycle analysis of patio heater show that the problem with the energy consumed and carbon foot print for the patio heater was during the use of the heater. A new materials can be selected to reduce the heat losses during the the use of the patio heater.

6. References

Alonso, E., Gregory, J., Field, F., Kirchain, R., (2007), Material availability and the supply chain: risks, effects, and responses'; Environmental Science and Technology, Vol 41, pp. 6649- 6656

Anees U Malik and P.C. Mayan Kutty, Corrosion and material selection in desalination plants, Presented to SWCC 0 & M Seminar, Al Jubail, April 1992.

Anees U Malik, Saleh A. Al-Fozan and Mohammad Al Romiahl , Relevance of corrosion research in the materials selection for desalination plants, Presented in Second Scientific Symposium on Maintenance Planning and Operations, King Saud University, Riyadh, 24-26 April, 1993

Ashby, M.F. (2005) "Materials Selection in Mechanical Design", 3rd edition, Butterworth-Heinemann, Oxford, UK, Chapter 16.

Ashby, M.F., Shercliff, H., and Cebon, D., (2007), "Materials: engineering, science, processing and design", Butterworth Heinemann, Oxford UK, Chapter 20.

Chapman, P.F. and Roberts, F. (1983), Metal resources and energy; Butterworth's Monographs in Materials, Butterworth and Co, ISBN 0-408-10801-0

Fiksel, J., Design for Envirnment, (2009), A guide to sustianble product development, McGraw Hill, ISBN 978-0-07-160556-4

Gabi, PE International, (2008), www.gabi-sofwtare.com

Graedel, T.E. (1998), Streamlined life cycle assessment, prentice Hall, ISBN 0-13-607425-1

Granta Design Limited, Cambridge, (2009) (www.grantadesign.com), CES EduPack User Guide

Hassan E. S. Fath and Mohamed A. Ismail, Enhancement of chemical cleaning and brine heater condensate, *Seventh International Water Technology Conference Egypt 1-3 April 2003*

Kyoto protocol, United Nations, Framework Convention on Climate Change, (1997), Document FCCC/CP 1997/7/ADD.1

Nafey, A. S., Fath, H. E. S., Mabrouk, A. A., Elzzeky, M. A., A new visual package for simulation of thermal desalination processes: development and verification, Eighth International Water Technology Conference, IWTC8 2004, Alexandria, Egypt

Wolfe, J.A. (1984), Mineral resources: a worls review, Chapman & Hall, ISBN 0-4122-5190-6

Part 3

Sustainable Manufacturing

5

The Kaolin Residue and Its Use for Production of Mullite Bodies

M.I. Brasileiro[1], A.W.B. Rodrigues[1], R.R. Menezes[2],
G.A. Neves[3] and L.N.L. Santana[3]
[1]*Federal University of Ceará, Campus Cariri, Cidade Universitária, Juazeiro do Norte*
[2]*Federal University of Paraíba, UFPB/PB, Cidade Universitária, João Pessoa*
[3]*Federal University of Campina Grande, Campina Grande,*
Brazil

1. Introduction

The continuous demand for higher productivity indices in the global competitive world has led to a fast decrease of the available natural resources and, at the same time, to the generation of a high volume of rejects or sub-products, most of them not directly recyclable.

The final report of the First National Conference on the Environment, 2003, emphasized the need for development of policies for environmental control and of restoration of the physical and biotic of the areas impacted by mining activities, encouraging the reuse, recycling and waste recovery and waste from mining (Neves et al., 2010).

Mineral extraction itself is a good example of reject production. Therefore, alternatives for recycling and/or reuse should be investigated and, where possible put into practice, since the waste is turned into serious problems with urban managing a complex and costly, considering the volume and mass uptake.

Mineral extraction and processing of kaolin is an example of generating a large volume of waste, such as kaolin ore and "industrial mineral" has many uses because of their color, low grain, low abrasiveness, chemical stability, specific form of constituent particles of clay minerals in addition to the specific rheological properties, appropriate in different fluid media (Lima, 2001; Almeida, 2006; Santana et al. 2007; Brasileiro, 2010). Thus, the kaolin is an industrial mineral that has its properties in the expansion of its use, and having as an ally of its low cost compared to most competing materials (Silva, 2001). Worldwide use is spread across various industrial sectors with emphasis on the paper (covering and filling) that consumes 45%, followed by ceramics (porcelain, white ceramic and refractory materials) 31% and the remaining 24% divided between ink , rubber, plastics and other (IBRAM, 2008).

In general, the mining of kaolin causes effects unwanted range that can be called externalities. These residues are in general, discharged on the environment, contaminating the ground and affecting the health of the population in the proximity of the producing regions. Some of these externalities: environmental change, land use conflicts, depreciation of surrounding properties, generation of degraded areas and urban traffic disturbances.

These externalities generate conflicts with the community, which usually originate when the implementation of the project because the developer was not informed about the expectations, desires and concerns of the community living near the mining company.

In this way, it comes growing the concern of environmentalist with the great amount of wastes produced in the kaolin industry. The main problems relate to the amount of waste generated and air pollution. The kaolin waste generated are simply piled up on land companies for processing, thus occupying a large volume. As the waste disposal sites are not correct, when dried, turn to dust and the wind, spread, polluting the air, roads, streams, etc.. Figures 1 and 2 show the site of deposition of the residue from kaolin, a company located in Paraiba, Brazil, after its recovery as well as the location is not suitable for deposition of the same, which is close to villages and rivers, respectively.

Fig. 1. View of the area used for disposal of waste generated during the processing of kaolin (Caulisa Company - City of Juazeirinho - Paraíba-Brazil)

Fig. 2. Location of improper waste disposal (Caulisa Company - City of Juazeirinho - Paraíba-Brazil)

Worldwide, millions of tons of kaolin waste, estimated at around 80% to 90% of the gross volume exploited, are produced each day in mining and mineral beneficiation.

One of aggravating of the kaolin residue deposit is that there is no exploitation, the waste is cumulative and cannot predict the future effects, since each batch received, twice its volume of waste is generated (SUDEMA, 2004). Another problem of this waste is that it is highly powdery, and its inhalation may cause lung disease and the skin contact causing dermatitis (Sakamoto, 2003).

The insertion of kaolin residue in the production cycle should represent an alternative option to recover these materials, which is interesting both in the environmental and health of the population next, as in economics. Recycling and reuse of kaolin residue are among the main alternatives in the quest for sustainable development, which aims to enable the economy of non-renewable raw materials and of energy, reducing the environmental impacts of waste in modern society.

The search for sustainable development makes the insertion of kaolin residue processing becomes the purposes of study in various areas of knowledge (Silva et al. 2001; Rocha, 2005, Menezes et al., 2007a, 2007b, 2007c; Resende, 2007), searching to minimize or even solve the environmental damage caused by these materials, either by optimization of the scanning method and storage or either by presentation of alternative re-use of waste as alternative raw material in various sectors.

As the processes of industrial processing, provide special features the kaolin for use in various industries, this means that the kaolin residue, has qualities that may be recycled or reused as a raw material for obtaining ceramic products.

The ceramic industry has demonstrated great potential for reuse of inorganic waste (Andreoli et al., 2002). The incorporation of waste from various industrial activities in ceramic products is an alternative technology to reduce environmental impacts and human adverse health effects caused by indiscriminate disposal of waste in nature (Özel et al. 2006; Pinatti et al, 2005).

Thus, the overall objective of this study was to investigate the viability of residual kaolin as an alternative raw material in obtaining mullite. Within this context, research will Facilitate the development of a technology practically nonexistent in Brazil and the world, which uses an alternative precursor, kaolin waste, which an alternative envisions the process with great potential for the synthesis of mullite. This would have positive social and economic consequences, both for generating industry of waste as for absorber industry, may promote a promising partnership between the privileged position of mullite, both domestically and abroad, and using waste as a raw material valued.

2. Potential of kaolin residue as secondary raw materials

The kaolin are those found in the place that formed by the action of weathering or hydrothermal action on rocks. The processing of this type of kaolin generates two types of waste. The waste first, generated by the wet process, which contains silica and others contaminants, resulting in amounts close to 70% by volume. The total volume of waste is very significant, since the improvement process has a yield of 30% of the total extracted from the deposit, ie, for each ton of raw material, less than one-third is taken advantage (these are the private data region, which can vary widely). The residue second is derived from the beneficiation of kaolin, separated in the previous step (about 30%), underwent screening in ABNT N° 200 mesh (0.074 mm). The material retained in the screening, called sludge, is the second residue, which was used in this study. This residue from processing of kaolin consists of minerals mica, quartz and feldspar, there is also the presence of kaolinite fraction whose amount is reduced depeding chiefly on the beneficiation process. The main product after sintering of kaolinite at high temperatures is the mullite (Lima et al., 2001; Chen et al., 2000). The ceramic products of kaolinitic have desirable properties of hardness, translucency and strength development of the mullite crystals, regardless of the effect, their number and their size and shape. The kaolin waste have appropriate characteristics so that may be used for making alumina-silica refractory materials, as they have in their composition the kaolinite as principal clay mineral, which when subjected to high temperatures produces mullite.

The high prices of materials ceramic of mullite base, cause there is a need to use alternative materials that are environmentally, low cost and with local availability.

Thus, there is increasingly a constant search by the various industries and economic development activities coupled with environmental preservation. In order to take advantage of the kaolin residue in the many sectors involved in the ceramic industry, several studies have been developed in recent decades. Among them can mention:

- Flores & Neves (1997), concluded that the methodology of alumina production through the use of tailings from the processing of kaolin is technically feasible, according to the parameters for the roasting, leaching and neutralization / crystallization adopted in his research.
- Miranda et al., (2000) studied the use of kaolin waste from industry benefited, with a view to the application as ceramic raw materials. Becoming evident that the possibility of using waste as raw material ceramic.
- Rocha (2005) studied the incorporation of the kaolin residue in the traces of mortar to be employed in construction activities. The results showed that traces of mortar containing kaolin residue are within the norms of ABNT, effecting thereby the feasibility of using this waste.
- Lima (2005) studied the potential of kaolin residue for use in plain concrete blocks with no structural function. The values of strength and absorption, obtained in all molded with the kaolin residue, are in accordance with the ABNT NBR NBR 7184/91 and 1211/91 respectively, thus demonstrating the feasibility of using the kaolin residue.
- Neves et al. (2006) studied the incorporation of the kaolin residue in ceramic bodies aiming at its use in the production of blocks for construction. The results showed the use of waste as raw material for production of ceramic blocks.
- Andrade et al. (2009), in their studies, feature and studied the processing of the waste from the benefit of kaolin industry with addition of plastic clay in formulations for ceramic tiles. Preliminary results showed that the residue studied can be considered as raw material of great potential for industrial floors and ceramic tile.

3. Using kaolin waste as raw material for obtaining mullite

Mullite is one of the most important aluminum silicate ceramic technology, the only stable intermediate compound in the system SiO_2-Al_2O_3, with the composition of $3Al_2O_3.2SiO_2$, corresponding to 71.8 wt%. The importance of technology, coupled with the occurrence rare in nature, emphasizes the importance of developing research and studies on the synthesis of mullite.

Because it is rare and almost nonexistent in nature, due to training conditions, ie high temperatures and low pressures (Schneider et al. 2008; Schneider & Komarneni, 2005, Vieira et al., 2007), present deposits are not enough to supply a growing market due to new applications found (Monteiro et al., 2004).

The mullite is perhaps one of the most important stages in both the traditional and advanced ceramics (Schneider et al. 2008; Juettner et al. 2007; Sola et al. 2006; Mazdiyasni et al., 1983). The exceptional scientific and technical importance of mullite can be explained on the basis of their characteristics, to be a stable crystalline phase from room temperature up to about 1880 °C at atmospheric pressure (Mileiko et al. 2009; Esharghawi et al. 2009; Schneider et al., 2008, Skoog et al., 1988), have in their properties the main reason for your application, such as low thermal expansion (20 / 200 °C = 4 x 10^{-6} K^{-1}), low conductivity thermal (k = 2.0 $Wm^{-1}K^{-1}$), low density (3.17 g/cm^3), low dielectric constant (ε = 6.5 at 1 MHz), excellent wear resistance and high resistance at temperatures suitable for use (Esharghawi et al. 2009; Schneider & Komarneni, 2005; Schneider, 1994; Meng et al. 1983; Bartscher et al. 1999; Viswabaskaran et al., 2003), and the fact that the raw materials for its production (eg alumina, silica, aluminum silicates, sheet silicates rich in Al_2O_3, clays, etc.) are widely found in nature (Schneider et al., 2008).

Various routes like sol-gel, co-precipitation, hydrothermal processes and procedures for chemical vapor deposition and sintering are already used for the synthesis of this mineral. However, these typically use chemical precursors and high temperatures.

The number of studies showing the important methods for obtaining mullite, has been growing in recent years. The composition, purity of reagents, the synthesis processes used dictate the properties of all ceramic materials. Therefore, the choice of a certain synthesis process of mullite is a key step to obtain mullite with desired properties and applications. In order to reduce production costs, it is necessary, the use of precursors more economically viable, highlighting the potential kaolin waste for this function.

Studies conducted by our research group (Menezes et al., 2009a, 2009b; Brasileiro et al. 2006, 2008; Santana et al., 2007) aimed at recycling the waste from the processing of kaolin, showed that the kaolin residue has high levels of SiO_2 and Al_2O_3, especially as a material of great potential for obtaining bodies that have properties for use in advanced ceramics, can be incorporated in ceramic formulations to obtain mullite.

This causes a secondary material to become a precursor suitable for obtaining a material that finds wide range of applications in conventional ceramics (porcelain, pottery, refractories, etc.) and advanced.

In general, it is necessary to observe the compositional variability of the kaolin residue with the quality of processing. Thus, each residue will have different potential for the synthesis, so that if the process is very efficient, or improved, there will be a reduction in the content of kaolinite, which likely will reduce the ability to produce mullite.

4. Methodology used

4.1 Processing of kaolin residue

The kaolin residue used in the development of the research was a result of primary processing of kaolin, extracted from the Borborema pegmatitic plain, located in the municipality of Juazeirinho-Paraíba - Brazil and donated by CAULISA Industry S/A.

This kaolin residue was dried at a temperature of 110 ° C for 24 hours until constant weight. Soon after, was mill a unbundle type and climbs past ABNT N° 100 (0.15 mm) then stored in plastic bags. According to Castro (2008), made an analysis of the results with all parameters of the material leached and dissolved, the residue from the beneficiation of kaolin, indicate that they are below the standardized (ABNT NBR 10004 (2004)), which fits this residue as type B class II waste - inert waste.

A small amount of residue powder kaolin, sieved in ABNT N° 200 (0.074 mm), was pressed in cylindrical shape, so this required by the equipment used in research. The sample of the kaolin residue, was analyzed by XRF in Shimadzu (EDX-900) thus determining its chemical composition (Table 1). Chemical analysis provides key data of great scientific and industrial use, although not a full evaluation of the mineralogical composition and physico-chemical and technological. The residue also sifted through sieve N° 200 ABNT (0.074 mm), was submitted to phase identification, carried out by X-ray diffraction, using an apparatus manufactured by Shimadzu, model XRD-6000 with CuKα radiation (40kV/40mA) at a speed of 2°/min from 15° to 60°.

4.2 Formulation of compositions for obtaining specimens

Being a careful and complex step in the production process of ceramic materials, the formulation of ceramic masses requires a prior knowledge of the raw materials that will be employed, seeking to establish a proper ratio of chemical raw materials for that particular phase is obtained. At this stage, took into account the chemical analysis of Alumina A1000SG, courtesy company Alcoa Industrial Chemicals Division, USA. The formulations (Table 2) of the ceramic bodies was performed with the aid of the REFORMIX 2.0, DEVELOPED BY THE Federal University of São Carlos – São Paulo – Brazil, being released data from the established results of chemical analysis of raw materials (kaolin residue), taking into account the desired stoichiometry of the mullite ($3Al_2O_3.2SiO_2$).

4.3 Process of forming and firing of the specimens

Formulated ceramic masses, they were mixed in ethyl alcohol (as described in the literature (Chen & Tuan, 2001)), together with the dispersing agent (PABA) and lubricant (oleic acid) in an alumina ball mill for 2h. After the mixing step, the masses were dried in an oven (110 ° C), unbundle and passed in sieve ABNT N°. 100 (0.15 mm). Then, the masses were moistened with 8% water, and thus stored for 24 hours for a better distribution of moisture. In step forming was used pressing process where the bodies of the test piece were conformed by uniaxial pressing of 35 MPa and dimensions (50 mm x 8 mm x 6 mm) in a hydraulic press (SCHWING / Siwa). The piece test bodies was dried in an oven at 110 °C for 24 hours and then subjected to the process of burning in a conventional electric oven (Maitec / Flyever / FE 50 rpm) at a temperature ranging from 1400 °C to 1600 °C/2h and heating rate of 5 °C / min in air atmosphere.

Since there is a growing interest in the use of microwave energy for processing materials at high temperatures, the specimens were subjected to sintering process in a domestic microwave oven for comparison. The oven had a frequency of 2.45 GHz, and the specimens were sintered in powers of 80 and 90% (output power of 1.44 kW and 1.62 kW, respectively) and irradiation times ranging from 10, 15, 20 and 25 min. The atmosphere of the microwave oven was the air. Figures 3 and 4 show the conventional and microwave oven, respectively, used in burning of the specimens.

Fig. 3. Conventional oven used in the burning of the specimens

Fig. 4. Domestic microwave oven used in research

4.4 Tests for mineralogical characterization of the specimens

After sintering of the bodies, it was ground in porcelain mortar and sifted through sieve N°. 200 ABNT (0.074 mm) and then submitted to phase identification, performed by diffraction X-rays, using an apparatus manufactured by Shimadzu, model XRD-6000 with CuKα radiation (40kV/40mA), pertaining to Department of Materials UFCG, the rate of 2°/min from 15° to 60°. The interpretation was performed by comparison with standards contained in Shimadzu software.

The bodies of the test piece were also subjected to microstructural characterization, performed in a scanning electron microscope, Phillips, model XL30FEG from the Department of Materials Engineering, Federal University of Paraíba. First analyzed the fracture surface and then the samples were polished and attacked with HF (10%) to remove the vitreous of the specimens and observed the morphology of mullite grains formed. For these tests, the samples were coated with a gold film using a vacuum evaporator.

5. Results

Preliminary tests made with the kaolin residue (Castro, 2008) showed that all parameters of the material are dissolved and leached below the standardized (ABNT NBR 10004 (2004)), which fits this residue as the residue class II type B - inert waste.

Figure 5 shows the diffractogram of the kaolin residue. From the XRD pattern verified that the kaolin residue has the following mineralogical phases: kaolinite characterized by interplanar distances of 7.07 and 3.56 Å, a small amount of mica characterized by 10.04 and 4.97 Å, and also quartz characterized by 2.66 and 2.08 Å.

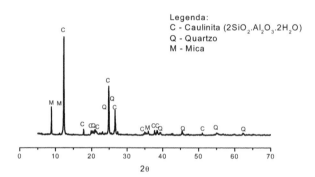

Fig. 5. XRD of the residue of kaolin

Figure 6 shows the test results for the particle size distribution of the kaolin residue used in this work.

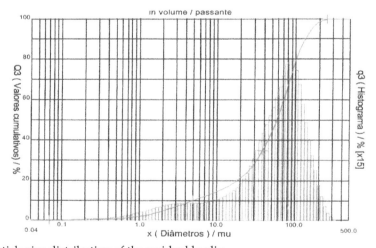

Fig. 6. Particle size distribution of the residual kaolin

The residue has a particle size of 54.35 m is a relatively broad size distribution, with D10 = 5mm, D50 = 58µm, D90 = 130µm, with a high fraction of particles with equivalent diameter greater than 40µm.

The chemical composition of the kaolin residue and alumina used in the course of this research is shown in Table 1.

The development of formulations used in this work was based on chemical analysis of raw materials.

It can be seen from Table 1 that the kaolin residue in its composition showed 52.68% of silicon oxide and 33.57% alumina oxide, therefore, due to high contents of SiO_2 and Al_2O_3, used to be used in obtaining mullite.

Alumina has analyzed high purity, 99.8% and was added to the kaolin residue in order to react with the glassy phase formed at high temperatures and thus form the secondary mullite.

The mullite, most commonly produced from the solid state reaction of powdered α-alumina and silica, is the mullite of composition 3:2, according to published data. The mullite present in this composition has been well studied with respect to the method of synthesis, thermal and mechanical properties (Yu et al., 1998).

Taking into account that the reaction $2SiO_2 \bullet Al_2O_3 \bullet 2H_2O + 2Al_2O_3 \rightarrow 3Al_2O_3 \bullet 2SiO_2 + 2H_2O$ occurs by a mixture of kaolinite and alumina during sintering, and assuming that the silica (SiO_2) in kaolinite reacts completely with the alumina (Al_2O_3) to form mullite (Chen et al., 2000), the kaolin residue that showed 52.68% of SiO_2, requires an excess of alumina for the reaction with silica, present in the residue, can result in mullite.

Constituents / Raw materials	SiO_2	Al_2O_3	Fe_2O_3	Na_2O	K_2O	TiO_2	CaO	MgO	*L.I
Kaolin residue	52.68	33.57	0.93	0.08	5.72	0.12	-----	-----	6.75
Alumina A1000 SG	0.03	99.8	0.02	0.06	------	-------	0.02	0.03	------

Table 1. Analysis of chemical constituents (in%) of the residue of kaolin and alumina. * L.I - Loss Ignation

Table 2 shows the compositions made of ceramic masses studied in this research.

Raw materials	Formulations (proporcion in mass weight %)				
	1	2	3	4	5
Kaolin residue	54	50	46	42	38
Alumina	46	50	54	58	62

Table 2. Formulations of mass (weight%)

Compositions were formulated with lower and higher levels of alumina, so that these formulations to stay within the stoichiometry of the mullite to be reached.

Table 3 presents the chemical composition (in%) of the compositions performed by XRF.

Sample	SiO₂	Al₂O₃	Fe₂O₃	CaO	NaO₂	P₂O₅	K₂O	MnO	TiO₂	P.F
Form. 1	28,46	64,01	0,51	0,01	0,07	---	3,09	---	---	3,64
Form. 2	26,35	66,68	0,47	0,01	0,07	---	2,86	---	---	3,37
Form. 3	24,25	69,33	0,44	0,01	0,07	---	2,63	---	---	3,10
Form. 4	22,14	71,98	0,40	0,01	0,07	---	2,40	---	---	2,83
Form. 5	20,03	74,63	0,37	0,01	0,07	---	2,17	---	---	2,56

Table 3. Chemical composition of masses by XRF (in%)

Analyzing the values in Table 3, it was found that the percentage of silica are in the range from 20 to 28% and the percentage of alumina in a range from 64 to 75%, indicating that the proportions of the oxides are approaching values for obtaining secondary mullite ($3Al_2O_3$. $2SiO_2$). These proportions of oxides also show that the formulation 2, was very close to the stoichiometry of the mullite to be reached 3:2 ($3Al_2O_3$. $2SiO_2$), the formulation 1 has less alumina and the formulas 3, 4 and 5, present with more alumina, values which grow gradually.

5.1 Microstructural features of the specimens sintered in a conventional oven

The XRD pattern of samples sintered in conventional oven at temperatures ranging from 1400 to 1600 °C, with a landing-burning 2h, are shown in Figures 7, 8, 9, 10 and 11 with the formulations 1, 2, 3, 4 and 5, respectively.

Through XRD, in all compositions, observed the presence of the following phases: mullite, quartz, cristobalite and alumina. At the temperature of 1400 °C the crystalline phases observed were mullite, quartz and alumina. At temperatures of 1450 °C to 1550 °C the crystalline phases were mullite, quartz, alumina and cristobalite. The appearance of cristobalite is due to the crystallization of vitreous silica. The intensity of the peaks of quartz and cristobalite, virtually disappear with increasing sintering temperature, while the intensities of the characteristic peaks of mullite increases with increasing sintering temperature.

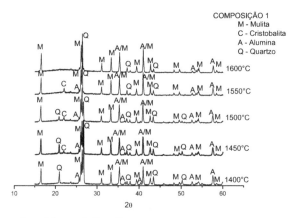

Fig. 7. XRD of samples of formulation 1

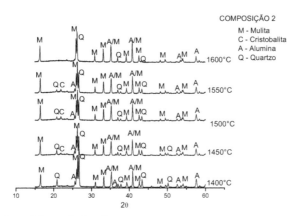

Fig. 8. XRD of samples of formulation 2

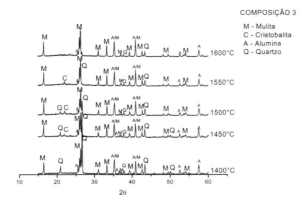

Fig. 9. XRD of samples of formulation 3

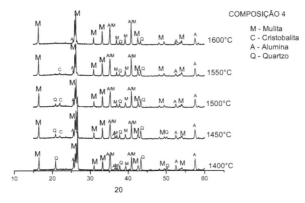

Fig. 10. XRD of samples of formulation 4

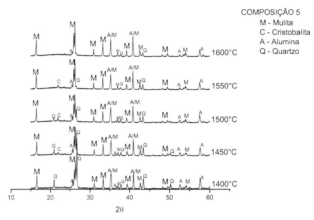

Fig. 11. XRD of samples of formulation 5

Figs 7 and 8, the which formulations have higher amounts of SiO_2, show alumina peaks a lesser degree when compared with other compositions, on the other hand the peaks of quartz present with greater intensity. According to Chen et al. (2000), the addition of Al_2O_3 can reduce the amount of glassy phase and increase the amount of mullite. This statement was confirmed by XRD patterns of formulations 3, 4 and 5.

For all compositions, observed that in 1600 °C, there was not a complete reaction of mulitization, became evident that even the formulation 2 (stoichiometrically correct) it is necessary a further increase in sintering temperature to occurs a complete mulitization.

Figures 12 (a and b), 13 (a and b), 14 (a and b), 15 (a and b) and 16 (a and b) show the microstructures of the formulations 1, 2, 3, 4 and 5 sintered in a conventional oven at temperatures of 1500 and 1550 °C, respectively.

Observed from the micrographs that the sintering provided the formation of a heterogeneous microstructure characterized by the formation of primary mullite, which is presented in the form of clusters of small crystals; secondary mullite, which is presented in

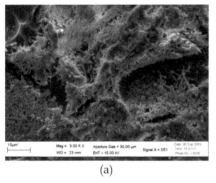
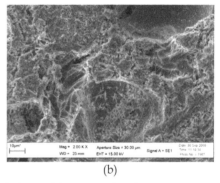

(a)   (b)

Fig. 12. (a and b) - Microstructure of formulation 1 sintered in conventional oven at temperatures of 1500 °C (a) and 1550 °C (b)

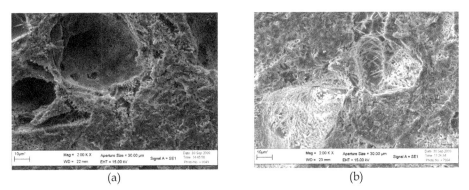

Fig.13. (a and b) - Microstructure of formulation 2 sintered in conventional oven at temperatures of 1500 °C (a) and 1550 °C (b)

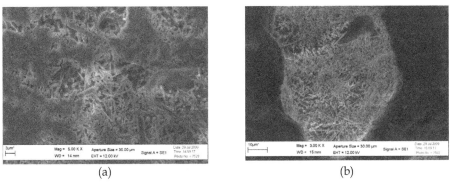

Fig. 14. (a and b) - Microstructure of formulation 3 sintered in conventional oven at temperatures of 1500 °C (a) and 1550 °C (b)

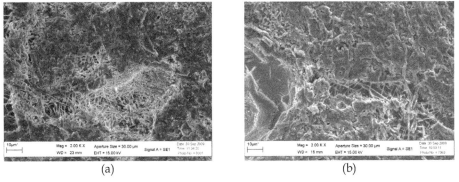

Fig. 15. (a and b) - Microstructure of formulation 4 sintered in conventional oven at temperatures of 1500 °C (a) and 1550 °C (b)

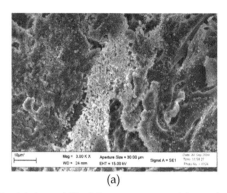
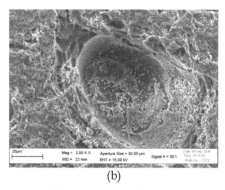

(a)   (b)

Fig. 16. (a and b) - Microstructure of formulation 5 sintered in conventional oven at temperatures of 1500 °C (a) and 1550 °C (b)

the form of long crystals and the presence pore. The formation of secondary mullite case of transient liquid phase, which happens to precipitation of crystals, ie the added alumina dissolves in the glass phase and mullite is precipitated. At 1500 °C was observed to obtain mullite, indicating that this system is a decrease in the eutectic point, with respect to the phase diagram composed of SiO_2 and Al_2O_3. Also in 1500 °C, observed the formation of both the primary mullite, which is presented in the form of clusters of small crystals, while the secondary mullite, is presented in the form of elongated crystals, like needles. The temperature rise from 1500 to 1550 °C provided the formation of a larger amount of secondary mullite and an increase in structure of the grains, making them more elongated, and the more intense the stretch as the temperature increased to 1550 °C. According to Iqbal & Lee (1999), the higher viscosity of the glassy phase favors the growth of crystals of mullite and also the continued growth in the size of the crystals indicates the possibility of primary mullite transformation of secondary mullite.

5.2 Microstructural features of the specimens sintered in a microwave oven

In domestic microwave oven used in the research, it was not possible to determine the temperature of use in the powers used, since it does not have provision for this. The XRD pattern of samples sintered in a microwave oven for 10, 15, 20 and 25 minutes are shown in Figures 18 (a e b), 19 (a e b), 20 (a e b), 21 (a e b) and 22 (a e b) with the formulations 1, 2 , 3, 4 and 5, respectively. These XRD patterns provide more information about the nature of synthesis from the microwave energy.

From the XRD patterns above can be seen that the formulation 1 (Fig. 17), subjected to 80% power and irradiation time of 15 min, began to show peaks of mullite into sharper focus when compared with other formulations. This formulation (see Table 3) presents a greater amount of SiO_2, which form in the sintering probably a greater amount of liquid phase, which would imply an acceleration in the reaction of formation of mullite. The formation of mullite was evident when the power used was 90% from 10 min. The peaks also show the presence of quartz and alumina.

Formulation 2 (Fig. 18), which has chemical composition closer to stoichiometry of mullite, sintered at 90% power and irradiation time of 20 min, presented in a manner relative to the intensity of the peaks, an amount greater formation of mullite, compared to formulation 1. Probably a slightly larger amount of alumina present in this sample, when compared to formulation 1, enabled a greater reaction between alumina and glassy phase for the formation of mullite.

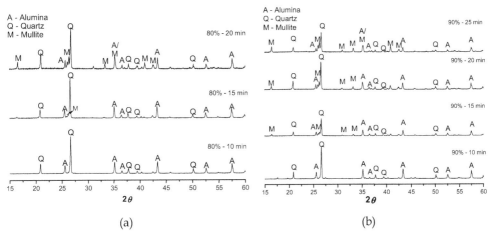

(a)　　　　　　　　　　　　　(b)

Fig. 17. (a and b) – XRD patterns of formulation 1 sintered in microwave at 80% power (a) (in times of 10, 15 and 20 minutes) and the power of 90% (b) (in times of 10, 15, 20 and 25 minutes)

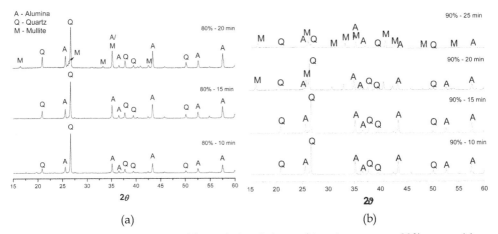

(a)　　　　　　　　　　　　　(b)

Fig. 18. (a and b) – XRD patterns of formulation 2 sintered in microwave at 80% power (a) (in times of 10, 15 and 20 minutes) and the power of 90% (b) (in times of 10, 15, 20 and 25 minutes)

Formulation 3 (Fig. 19), with amount of alumina most than formulation 2, did not show the formation of mullite in equal proportions, when compared with formulations 1 and 2 when sintered in the power of 90% and time 20 min, what can this associated with a likely absence of liquid phase.

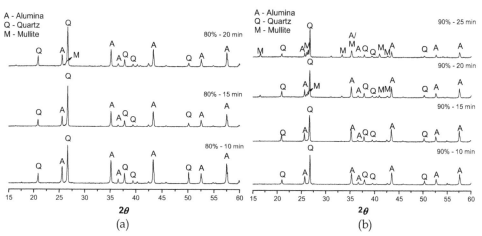

Fig. 19. (a and b) – XRD patterns of formulation 3 sintered in microwave at 80% power (a) (in times of 10, 15 and 20 minutes) and the power of 90% (b) (in times of 10, 15, 20 and 25 minutes)

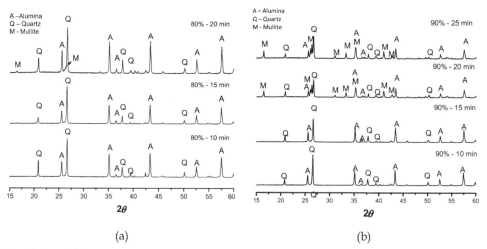

Fig. 20 (a and b) – XRD patterns of formulation 4 sintered in microwave at 80% power (a) (in times of 10, 15 and 20 minutes) and the power of 90% (b) (in times of 10, 15, 20 and 25 minutes)

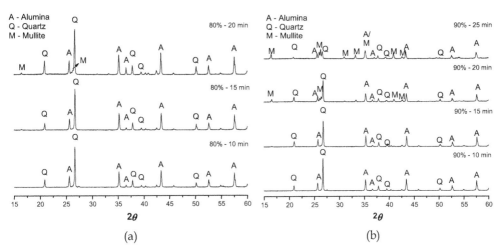

Fig. 21. (a and b) – XRD patterns of formulation 5 sintered in microwave at 80% power (a) (in times of 10, 15 and 20 minutes) and the power of 90% (b) (in times of 10, 15, 20 and 25 minutes)

The formulations 4 and 5 had similar behaviors when sintered at a power of 90% and times 20 min. The XRD patterns showed peaks of mullite formation, indicating that there was a reaction between the liquid phase and alumina, and that perhaps this composition range, the reaction went to the side of the eutectic phase diagram of the system Al_2O_3-SiO_2 favoring the formation of mullite.

In general, the formulations subjected to 90% power and times of 25 minutes, increased the peaks corresponding to formation of mullite and on the other hand, there was a decrease in the number of peaks of quartz in the amount of alumina peaks.

Analyzing powers and times used, it can be said that the power of 90% and 25 min time favored the formation and growth of the mullite peaks at all compositions, and these processing parameters were still insufficient for a total reaction between alumina and silica to form mullite, since the XRD patterns indicate the presence of quartz and alumina. Figures 22 (a, b and c), 23 (a, b and c), 24 (a, b and c), 25 (a, b and c) and 26 (a, b and c) show the microstructures of the formulations 1, 2, 3, 4 and 5, sintered in a microwave oven with times of 20 and 25 minutes and the power of 80% and 90%.

It can be seen from the figures (22-26), the main phases found in the samples are mullite primary, secondary mullite, quartz particles and pores. The quartz grains are surrounded by cracking due to the large difference between the coefficient of thermal expansion of quartz ($\alpha \approx 23 \times 10^{-6} K^{-1}$) and glassy phase ($\alpha \approx 3 \times 10^{-6} K^{-1}$). The primary mullite is present in the form of compact aggregates of small crystals. It is also observed that the sintered samples, in the power of 80% and time 20 minutes, show the formation of needles, proving the acquisition of secondary mullite. The process of formation of secondary mullite needles continues and becomes even more evident when the samples are sintered in the power of 90% and time 20 minutes. The samples sintered in 90% power and time of 25 minutes revealed a significant degree of recrystallization and a reduction in the appearance of needles. Similar results of

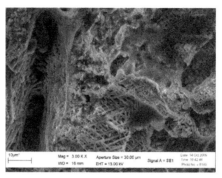

(a) 20 min in the power of the 80%

(b) 20 min in the Power of the 90%

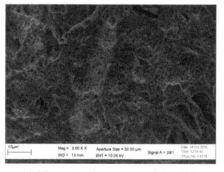

(c) 25 min in the power of the 90%

Fig. 22. (a, b and c) - Microstructure of formulation 1 sintered in microwave power of 80% and 90% and times of 20 and 25 minutes

(a) 20 min in the Power of the 80%

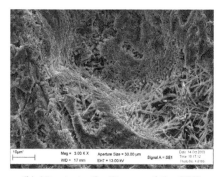

(b) 20 min in the Power of the 90%

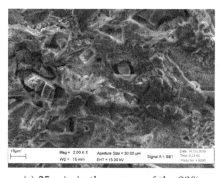

(c) 25 min in the power of the 90%

Fig. 23. (a, b and c) - Microstructure of formulation 2 sintered in microwave power of 80% and 90% and times of 20 and 25 minutes

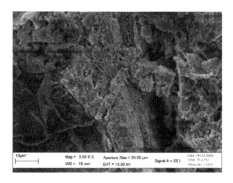

(a) 20 min in the power of the 80%

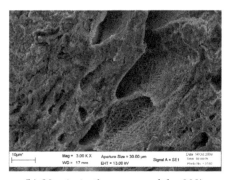

(b) 20 min in the power of the 90%

Fig. 24. (a, b and c) - Microstructure of formulation 3 sintered in microwave power of 80% and 90% and times of 20 and 25 minutes

recrystallization were shown by Panneerselvam & Rao (2003) in their studies. It was also possible to observe that the formation of secondary mullite occurs from large "gaps" or holes, probably due to the presence of the glassy phase.

By making a comparative analysis of the XRD diffractograms and microstructure by SEM, it was found that the formulations 1 and 2, sintered in the power of 80% and time 20 min, showed peaks of mullite into sharper focus when compared to formulations 3, 4 and 5, noting that the amount of glassy phase present in these formulations, accelerated the reaction sintering with alumina to form mullite. This acceleration may have been caused due to the microwave effect, in which the microwave field results in drivers force that reduce the barriers of reaction very effectively (Panneerselvam & Rao, 2003). On the other hand, the compositions with larger amounts of alumina needed to be submitted to the power and longer time to show the formation of secondary mullite.

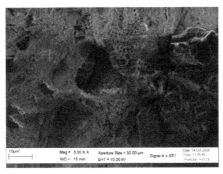

(a) 20 min in the power of the 80%

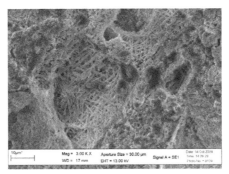

(b) 20 min in the power of the 90%

(c) 25 min in the power of the 90%

Fig. 25. (a, b and c) - Microstructure of formulation 4 sintered in microwave power of 80% and 90% and times of 20 and 25 minutes

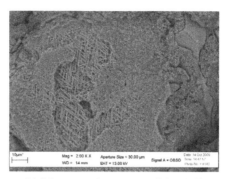

(a) 20 min in the power of the 80%

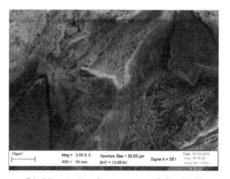

(b) 20 min in the power of the 90%

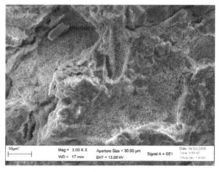

(c) 25 min in the power of the 90%

Fig. 26. (a, b and c) - Microstructure of formulation 5 sintered in microwave power of 80% and 90% and times of 20 and 25 minutes

5.3 Comparison of results of the specimens sintered in microwave and conventional oven

From the results obtained by the XRD diffractograms, it was observed that in just 20 min of sintering in a microwave oven, was the formation of secondary mullite. Time of 280 min, for example, were required in conventional sintering, when the temperature was 1400 ° C for the formation of secondary mullite.

Ebadzadeh et al. (2009), stated in his studies that the process with microwave led to reduction 120 °C reduction in sintering temperature of specimens. Panneerselvam & Rao (2003), stated that the maximum temperature of reaction, in obtaining mullite in a microwave oven was only 1312 °C, which is much lower (300 - 400 °C), the temperatures reported in conventional oven .

This fact focuses on the possible occurrence of a genuine effect of microwaves. According Panneerselvam & Rao (2003), the microwave field is very intense in the interfaces, surfaces and cavities during irradiation. The microwave field results in drivers force that reduce the barriers of reaction very effectively. Thus, during the formation of secondary mullite, in particular, the effect of driving force can be responsible for the rapid diffusion of Al^{+3} ions in layers of SiO_2.

By comparing the structure microstructure of the sintered samples in a microwave oven (Fig. 22 to 26) and in conventional oven (Fig. 12 to 16), can be seen that the microwave sintering provided the formation of secondary mullite needles more elongated and intertwined with each other. In all micrographs presented, it was not possible to measure the exact size of grains of single crystals due to aggregation and interconnectivity of the grains of primary and secondary mullite.

6. Conclusion

It was proved that after the study using kaolin waste in mixtures with alumina, it was concluded that these residues had characteristics suitable for the proposed use.

In this research the formation of secondary mullite is processed very quickly from the microwave, which might make this procedure satisfactory with regard to the economic aspect.

Factors such as power, time and temperature used was insufficient for complete reaction between alumina and silica to form mullite, however, the results indicate that mullite powders with better properties can be obtained from the microwave, the optimization of processing parameters.

Unable to determine the energy used during the reaction between alumina and silica to form mullite, both by microwave and by conventional oven, where such ovens did not have equipment for such analysis.

The results show the importance of studying the kaolin residue in obtaining mullite, conclude that, in addition to existing technical possibilities, one must highlight the socio-economic viability of the process of obtaining mullite from kaolin residue, where large volumes of this are accumulated indiscriminately.

The recycling of materials, represented here in this research by reusing the kaolin residue, is more than one action to try to save the planet, is one of the most important ways to ensure the development of peoples from the rational use of materials.

7. Acknowledgements

The authors thank CAPES and CNPq for financial support. Processes 479674/2007-8 and 307068/2007-2.

8. References

ABNT – ASSOCIAÇÃO BRASILEIRA DE NORMAS TÉCNICAS. NBR 10004: Resíduos Sólidos – Classificação, Rio de Janeiro, 2004.

Almeida, R.R. (2006). Reciclagem de resíduo de caulim e granito para produção de blocos e telhas. *Dissertação de Mestrado.* Campina Grande: CCT/UFCG, Coordenação de Pós-Graduação em Engenharia Civil, pp. 116..

Andrade, F.L.F., Varela, M.L., Dutra, R.P.S., Nascimento, R.M., Melo, D.M.A., Paskocimas, C.A. (2009). Avaliação da Potencialidade de Uso do Resíduo Proveniente da Indústria de Beneficiamento do Caulim na Produção de Piso Cerâmico. *Cerâmica Industrial,* 14, 1, pp.41-45.

Andreola, F., Barbieri, L., Corradi, A., Lancellotti, I., Manfredini, T. (2002). Utilisation of municipal incinerator grate slag for manufacturing porcelainized stoneware tiles manufacturing. *Journal European Ceramic Society* 22, pp.1457-.

Bartsch M., Saruhan B., Schmucker M., Schneider H. (1999). Novel low-temperature processing route of dense mullite ceramics by reaction sintering of amorphous SiO_2-coated γ-alumina particle nanocomposites. *Journal American Ceramic Society,* 74, pp. 2448-2452.

Brasileiro, M.I. et.al., (2006). Mullite Preparation from Kaolin Residue. *Material Science Forum,* v.530-531, pp. 625-630.

Brasileiro, M.I. et.al. (2008). Use of Kaolin Processing Waste for the Production of Mullite Bodies. *Material Science Forum,* v.591-593, pp. 799-804.

Brasileiro, M.I. (2010). Síntese de mulita por microondas utilizando composições contendo resíduo de caulim. *Tese de Doutorado.* Campina Grande, PB. Doutorado em engenharia de Processo CCT/UFCG.. pp. 112.

Castro, W.A.M.(2008). Incorporação de resíduo de caulim em argamassas para uso na construção civil. *Dissertação de Mestrado.* Campina grande: CCT/UFCG, Programa de Pós-Graduação em Ciência e Engenharia de Materiais. 98p.

Chen, C. Y., Lan, G.S., Tuan, W. H. (2000). Microstructural evolution of mullite during the sintering of kaolin powder compacts. *Ceramic International,* 26, p. 715-720, 2000.

Chen, C. Y.,Tuan W. H. (2001). The processing of kaolin powder compact. *Ceramic International,* 27,pp. 795-800.

Esharghawi, A., Penot, C., Nardou, F. (2009). Contribuition to porous mullite synthesis from clays by adding Al and Mg powders. *Journal European Ceramic Society,* 29, pp.31-38.

Flores, S.M.P., Neves, R.F. (1997). Alumina para utilização cerâmica, obtida a partir do rejeito de beneficiamento de caulim. 41° Congresso Brasileiro de Cerâmica, São Paulo. Available from:

http://www.scielo.br/scielo.php?script=sci_pdf&pid=S0366-9131997000400005&Ing=es&nrm=iso&tlng=pt.

IBRAM (2008) - Instituto Mineral Brasileiro – *Panorama Mineral Brasileiro 2008*. Disponível em: www.ibram.org.br. Available from: 20 de junho de 2011.

Iqbal, Y., Lee, W.E. (1999). Fired porcelain microstructures revisited. *Journal American Ceramic Society*, 82, 12, pp. 3584-3590.

Juettner, T., Moertel, H., Svinka, V., Svinka, R. (2007). Structure of kaolinite-alumina based foam ceramics for high temperature applications. *Journal European Ceramic Society*, 27, pp. 1435-1441.

Lima, F.T.; Gomes, J.; Neves, G.A.; Lira, H.L. (2001). Utilização do resíduo industrial resultantes do beneficiamento de caulim para fabricação de revestimentos cerâmicos. In: *45º Congresso Brasileiro de Cerâmica*, 2001, Florianópolis - SC. Anais. Associação Brasileira de Cerâmica, Florianópolis-SC, pp. 1501-1512.

Lima, M. S. (2005). Utilização do Resíduo de Caulim para uso em Blocos de Concreto sem Função Estrutural. *Dissertação de Mestrado*. (Mestrado em Engenharia Civil e Ambiental)-Universidade Federal de Campina Grande. Campina Grande, PB, 2005.

Mazdiyasni, K. S., Mah. (1983). Mechanical Properties of Mullite. *Journal American Ceramic Society*, 66, 10, pp. 699-704.

Menezes, R.R., De Almeida, R.R., Santana, L.N.L., Neves, G.A., Lira, H.L., Ferreira, H.C. (2007a). Análise da co-utilização do resíduo do beneficiamento do caulim e serragem de granito para produção de blocos e telhas cerâmicos. *Cerâmica*, v.53, p. 192-199.

Menezes, R.R.; Oliveira, M.F.; Santana, L.N. L.; Neves, G. A.; Ferreira, H.C. (2007b). Utilização do resíduo do beneficiamento do caulim para a produção de corpos mulíticos. *Cerâmica*, v.53, n°328, p.388-395.

Menezes, R.R., Souto, P.M., Kiminami, R.H.G.A. (2007c). Microwave hybrid fast sintering of porcelain bodies. Journal Materials Process Tecnic. 190, p. 223-229.

Menezes, R. R. ; Farias, F. F. ; Oliveira, M. F. ; Santana, L. N. L. ; Neves, G. A. ; Lira, H. L. ; Ferreira, H. C. (2009a). Kaolin processing waste applied in the manufacturing of ceramic tiles and mullite bodies. *Waste Management & Research (ISWA)*, v. 27, pp. 78-86.

Menezes, R. R.; Brasileiro, M. I. ; Gonçalves, W. P. ; Santana, L. N. L.; Neves, G. A. ; Ferreira, H. S.; Ferreira, H. C. (2009b). Statistical design for recycling kaolin processing waste in the manufacturing of mullite-based ceramics. *Materials Residue*, v. 12, pp. 201-209.

Meng, G., Huggins, R. A. (1993). New Chemical Method for Preparation of both Pure and Doped Mulite. *Materials Residue Bulletin*, v.18 (5),pp. 581-588.

Mileiko, S.T., Serebryakov, A.V., Kiiko, V.M., Kolchin, A.A., Kurlov, V.N., Novokhatskaya, N.I. (2009). Single crystalline mullite fibres obtained by the internal crystallisation method: Microstructure and creep resistance. *Journal European Ceramic Society*, 29, pp. 337-345.

Miranda, E. A. P., Neves, G. A., Lira, H. L. (2000). Reciclagem de Resíduos de Caulim e Granito para Uso Como Matéria-Prima Cerâmica. *Anais do 45º Congresso Brasileiro de Cerâmica*, Florianópolis – SC- Brasil.

Monteiro, R.R., Sabioni, A.C.S., da Costa, A, G.M. (2004). Preparação de mulita a partir do mineral topázio. Cerâmica, v.50, pp.318-323.

Neves, G.A., Almeida, R.R., Santana, L.N.L., Lira, H.L., Ferreira, H.C. (2006). Blocos cerâmicos utilizando resíduo de caulim. *Anais do 50º Congresso Brasileiro de Cerâmica* – Blumenau – SC – Brasil.

Neves, Gelmires de Araújo et al., (2010). *Resíduos industriais na construção de habitações de interesse social*. 1º Edition. Gráfica e Editora Agenda. Campina Grande – Paraíba – Brazil.

Ozel, E., Turan, S., Çoruh, S., Ergun, O. N. (2006). *Waste Management and Research* , v.24, pp.125.

Panneerselvam, M. & Rao, K.J. (2003). Novel Microwave Method for the Synthesis and Sintering of Mullite from Kaolinite. *Chemical Materials*, v.15, pp.2247-2252.

Pinatti, D. G., Conte, R. A., Borlini, M. C., Santos, B. C., Oliveira, I., Vieira, C. M. F., Monteiro, S. N. (2005). Incorporation of the ash from cellulignin into vitrified ceramic tiles. *Journal European Ceramic Society*, 26, p.305.

Rezende, M.L.S. (2007). Estudo de viabilidade técnica da utilização do resíduo de caulim em blocos de vedação. *Dissertação de Mestrado*. (Mestrado em Engenharia Agrícola) – Coordenação de Pós-Graduação em Engenharia Agrícola,Universidade Federal de Campina Grande - Campina Grande.

Rocha, A. K. A. (2005). Incorporação de Resíduo de Caulim em Argamassa de Alvenaria. *Dissertação de Mestrado*. (Mestrado em Engenharia Civil e Ambiental)-Universidade Federal de Campina Grande. Campina Grande, PB. pp.95

Sakamoto, L. (2003). *Os homens-tatu do sertão*. Sertão potiguar, Rio Grande do Norte. Available from: http://www.reporterbrasil.com.br/caulim/iframe.php. 10/01/2010.

Santana, L. N. L. et.al. (2007). Influência das matérias-primas em corpos cerâmicos contendo resíduo de caulim. *Anais do 51º Congresso Brasileiro de Cerâmica de* 3 a 6 de junho 2007 – Salvador – Bahia - Brazil - pp. 1-12.

Schneider, H., Okada, K., Pask, J. (1994). Mullite Synthesis. *Mullite and Mullite Ceramics*. Publisher: John Wiley & Sons Ltda. Chichester: pp. 105-119.

Schneider, H., Komanerni, S. *Mullite*. Federal Republic of Germany, 2005.

Schneider, H., Schreuer, J., Hildmann, B. Structure and properties of mullite – A review. *Journal European Ceramic Society*, 28, p.329-344, 2008.

Scliar, C. (2003). *Agenda 21 e o setor mineral*. Brasília: Secretaria de Políticas para o Desenvolvimento Sustentável – MMA.

Silva, A.C., Vidal, M., Pereira, M.G. (2001). Impactos ambientais causados pela mineração e beneficiamento de caulim. *Rem: Revista Escola de Minas*, v.54, n.2, pp. 133-136.

Skoog, A. J., Moore, R. E. (1988). Refractory of the past for the future mullite and its use as a bonding phase. *Journal American Ceramic Society Bulletin,* 67, pp. 1180-1185.

Sola, E.R., Torres, F.J., Alarcón, J. (2006). Thermal evolution and structural study of 2:1 mullite from monophasic gels. *Journal European Ceramic Society*, 26, pp. 2279-2284.

SUDEMA (2004) – Superintendência de Administração do Meio Ambiente. *Inventário de resíduos sólidos industriais do Estado da Paraíba*. João Pessoa.

Vieira, S.C., Ramos, A.S., Vieira, M.T. (2007). Mullitization kinetics from silica and alumina-rich wastes. *Ceramic international*, v. 33, pp. 59-66.

Viswabaskaran, V., Gnanam, F. D., Balasubramanian, M. (2003). Mullitisation behaviour of calcined clay-alumina mixtures. *Ceramic International*, v.29, pp.561 – 571.

Yu, J., Shi, J.L., Yuan, O.M., Yang, Z.F., Chen, Y.R. (1998). Mullitisation and Densification of alumina-rich aluminosilicate diphasic gel. *Journal Materials Scince Letters*, v.17, pp.1691-1693.

6

Assessing Long Term Sustainability of Global Supply of Natural Resources and Materials

K.V. Ragnarsdóttir[1], H.U. Sverdrup[2] and D. Koca[2]
[1]*Faculty of Earth Sciences, University of Iceland, Reykjavik*
[2]*Applied Systems Analysis & System Dynamics Group, Department of Chemical Engineering, Lund University, Lund*
[1]*Iceland*
[2]*Sweden*

1. Introduction

The human population has grown exponentially over the past century and is expected to increase to nine or ten billion by the year 2050 (Evans, 1998). This growth has been accompanied by an increasing rate of consumption of natural resources (Brown & Kane, 1994, Brown, 2009a,b). On several key resources, the use of materials and energy has increased faster than the population growth alone. At present, humans are challenging planetary boundaries and capacities (Humphreys et al., 2003, Rockström et al., 2009). For many fossil resources (energy, most metals and key elements), the rate of extraction is now so high that it can only with difficulty be further increased (Hubbert, 1956, Pogue & Hill, 1956, Ehrlich et al., 1992, Smil, 2001, 2002, Fillipelli, 2002, 2008, Greene et al., 2003, Arleklett, 2003, 2005, Hirsch et al., 2005, Gordon et al., 2006, Heinberg, 2007, Zittel & Schindler, 2007, Roskill Information Services, 2007a,b,c,d, 2008, 2009a,b, 2010a,b,c, 2011, Strahan, 2007, 2008, Ragnarsdottir et al., 2011, Sverdrup & Ragnarsdottir, 2011). In many cases, known resources are dwindling, because prospecting cannot find more. There have been several earlier warnings about the prospect of upcoming future material scarcity (Forrester, 1971, Meadows et al., 1972, 1992, 2004, Graedel & Allenby, 1995), though these have been seen as "interesting", but have generally been shrugged off as academic studies. In the years after world war II, there has been a redefinition of success and wealth to imply increased consumption and material through-put (Friedman, 1962, Friedman & Friedman, 1980, Jackson, 2009). This success, reported as gross national product (GDP), has been adopted by most leaders of the world as a generic measure of success (growth), leading to enormous flows of materials, and as a result, waste. Fossil fuels are arguably the most essential modern commodity that may become scarce during the coming decades (Hubbert, 1966, 1972, 1982, Hirsh, 1992, Graedel et al., 1995, 2002, 2004), but rare minerals and metals, used, for example, in mobile phones, are also not in unlimited supply (Cohen, 2007, Ragnarsdottir, 2008). New technologies, such as transistors, pin-head capacitors, compound semiconductors, flat-screen liquid-crystal displays, light emitting diodes, electric car batteries, miniature magnets and thin-film solar cells therefore need to be developed according to the long-term availability of their key material ingredients.

There are some important facts we need to keep in mind at all times when considering many of our essential resources. Most of them represent inheritance from past geological times, and the amounts regenerated per year are vanishingly small compared to our present use. The global commons has only a one-time allotment for all ages and generations. The amounts are finite, and if we use them all now, then we deprive future generations of many possibilities to support them selves (Norgaard & Horworth, 1991, Ainsworth & Sumaila, 2003, Heinberg, 2007, Brown, 2009b). At present, for every ton of natural resources we remove and waste irreversibly, there will be that amount permanently less in available stock on Earth for future generations.

From limits to accessibility

Around 50 different metals and elements are necessary to produce cars, computer chips, flat-screen TVs, DVD players, mobile-phone screens, hybrid cars, compact batteries, miniature machinery and cameras. For computer chips, this number has increased from around 10 different metals in 1980 to more than 40 metals and elements today. The concentrated ore deposits of these metals that can be easily tapped through mining are finite, even if - overall, the metals are in sufficient supply in the Earth's crust. But are we running out of metals that lie at the heart of our technological society? At an international conference in 2008 (Hall, 2008, Williams, 2008), experts from numerous geological surveys and mining companies claimed that many resources are yet to be mined, arguing that the key is to mine more deeply (up to 2 km) for lower-grade ores and to exploit the ocean floor (up to 7 km below sea level) (Ragnarsdottir, 2008). But deeper mining and refining of lower-grade ores will require more energy (Hall et al., 2001, 2008, Roskill Information Services, 2007a,b,c,d, 2008, 2009a,b, 2010a,b,c, 2011, USGS, 2008), another precious resource. And it remains as a fact, that these views are more wishful thinking than based in any reality. Very few mines go below the 1,500 meter mark in mountains, harvesting the deep seas has not been notably successful, and there is limited technology available at present to undertake such mining. As elements go scarce and hence, expensive, mining will move to also extract from low-grade and ultra-low grade reserves. However, there are limitations to that practice. At some point, the material and energy expenditure exceeds the use that can be obtained out of the extracted resource, thus it does not pay off – usually referred to as EROI (Energy Return On Investment; Hall, 2008). One usual argument for example is that there is theoretically enough phosphorus on earth for all times. However, when we consider that what is available when phosphorus has been dissipated or lost to the sea, the energy and material expenditure to have it extracted is not economically viable. Many new technologies may end up costing too much in metals and energy to build for a mass production scale so that, at least some of them, will never cover the EROI. Even though such solutions may appear as a good idea, they are unsustainable.

Various methods have been developed to analyse material flows through socitety. **Material flow analysis**, MFA (or substance flow analysis, SFA) is an analytical method for quantifying flows and stocks of meterials or substances in a well-defined system. MFA is an important tool to quantify the physical consequences of human activities and needs. It is used in the field of Industrial Ecology for different spatical and temporal scales. Examples include accounting of material use by different societies, and development of strategies for improving the matrerial flow systems as material flow management (Graedel & Allenby, 1995, Brunner and Rechberger, 2003). **Life cycle assessment** (LCA or life cycle analysis, ecobalance, and cradle to grave analysis) is a technique to assess environmental impacts associated with all the stages of a product's life from cradle to grave. The LCA includes compilation of an inventory of relevant energy and material inputs and environmental releases; the evaluation of the potential impacts associated with identified inputs and releases; and interpretation of results to help make informed decisions (e.g. Guinée, 2002). In the life cycle of a product the life cycle inventory as part of live cycle assessment can be considered an MFA as it involves system definition and balances. Cradle to cradle analysis is a specific kind of cradle to grave assessment, where the end of life disposal step for the product is a recycling process.

In this chapter, we use **systems analysis** and **system dynamics** (Forrester, 1961) as well as **burnoff time** and **Hubbert curve** representation to assess long term sustainability of global supply of natural resources and materials. **Systems Analysis** deals with analysis of complex systems by creating conceptual model structures with the help of **Causal Loop Diagrams** (CLD). CLDs make clear the cause and effect relationships and the feedbacks between different components in a system. **System Dynamics** is a methodology used to understand the behavior of complex systems over time. It deals with internal feedback loops and time delays that affect the behavior of the entire system. With the help of systems dynamics the conceptual model structures are transferred into dynamic numerical models, which can be then used as decision support tools, enabling the user to generate different scenarios and analyze the associated simulation results. Systems analsyis and system dynamics provides a deep insight in identifying interdependencies and feedback processes of dynamic stocks and flows of materiels, population dynamics and recycling rates. Neither MFA nor LCA encompasses these imporant components of systems analysis.

2. Methods of assessment

We use three different types of methods in order to estimate the time horizon of a raw material or metal resource:

1.   **Burn-off time:**

We define burn-off time as known mineable reserves divided by the estimated average annual mining rate. The formula is given as:

$$\text{Burn-off time} = \text{reserves/mining rate [yrs]} \tag{0}$$

2.   **Hubbert's peak resource estimate:**

An oil engineer at Shell Oil Corporation, developed what is referred to as the "Hubbert curve" (Hubbert, 1956, 1966, 1972, 1982) in order to predict the lifetime of oil wells and oil fields. He showed, using observed production data for oil wells as well as metal and phosphate mining, that all finite resource exploitation follows a distinct pattern of the Hubbert curve[1] (Figure 1). The shape of the Hubbert curve has a scientific explanation deriving from the nature of a finite resource, as well as the fact that Hubbert could verify his model on field data several times over. The Hubbert curve is defined by:

$$M = M_{max} / (1 + ae^{-bt}) \tag{1}$$

---

[1] http://en.wikipedia.org/wiki/Hubbert_peak_theory

where M_{max} is the total resource available (ultimate recovery of crude oil), M the cumulative production, and the coefficients a and b are constants.

We adapted the Hubbert curve and define the annual production as:

$$P = 2 P_{max} / (1 + \cosh(b(t-t_{max}))) \qquad (2)$$

where P_{max} is the maximum production rate, P is the production at time t, t_{max} is the time of the peak, and the coefficient b is the curve shape constant. Available history for the source, the size of the reserve and $1/3$ of the production curve is enough to set the a and b coefficients (Cavallo, 2004).

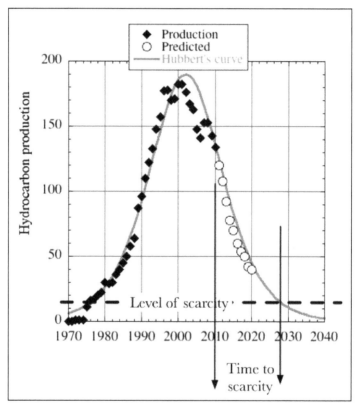

Fig. 1. The extraction pattern for hydrocarbon follows a symmetrical curve common to extraction of all resources. This can be used with observations of production rates to estimate the time to scarcity as shown with an analysis of the Norwegian oil production. The diagram suggests that the time to scarcity for oil produced in Norway is about 15 years. By 2040, the Norwegian age of oil production from Norwegian oilfields will be over.

3. Systems analysis and system dynamics:

Systems analysis with the help of causal loop diagrams is essential for gaining insights into the world metal supply system. With Figure 2, we intend to show that despite the Hubbert

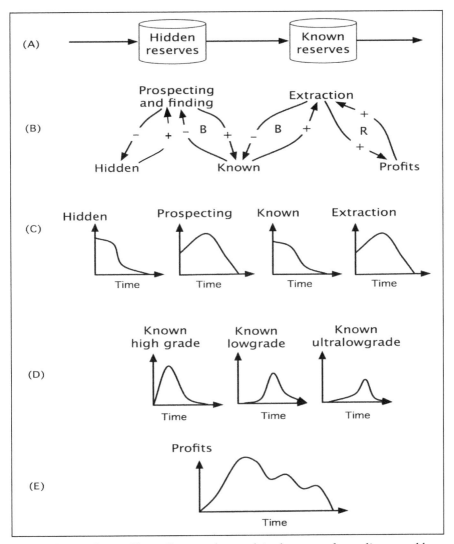

Fig. 2. Flow chart and causal loop diagram that explain the curve-shape discovered by Hubbert (1956, 1982). The system is based on two stocks, one known resource, backed by a hidden resource that can be found by prospecting for it and converting "hidden" to "known" (A). As "hidden" dwindles, "known" does get replenished until "hidden" is exhausted. As the extraction loop exhausts the "known", it gets backfilled, overlying several individual "rise and fall" curves to yield the typical Hubbert curve. The causal loop diagram is shown in (B). The resulting behavior of the components is depicted in (C). If we consider three types of resources in the system; high-grade, low-grade and ultralow-grade, we get three individual peak behavior curves (D), which may be overlaid to the final total production, expressed as profits (E). The results of using the Hubbert curve and using systems analysis and causal loop diagramming yield similar results for time to scarcity.

curve being empirically determined, there is a mechanistic explanation for it. In Figure 2, a flow chart (A) is shown with the corresponding causal loop diagram (B). The system is based on two stocks, one known resource, backed by a hidden resource that can be found by prospecting and finding more resource, converting "hidden" to "known". As the hidden resource, which is finite, dwindles, the "known" resource gets replenished for a short time until the "hidden" resource is exhausted. Thus as the extraction loop exhausts the "known" resource, it gets backfilled a few times by prospecting that brings more "hidden" resource over to "known" (B, C), overlying several individual "rise and fall" curves (D) to yield the typical Hubbert curve (E). The system is driven from the profit side, as mining leads to profits and more profits drives more mining, whereas the exhaustion of the finite stocks terminates it. The diagrams show the relationship between the parameter at the arrow-head over time. The diagrams depict what happens with new resources becoming known after prospecting (D), the bottom diagram (E) shows the sum of all the small diagrams, depicting extraction as a function of time.

The basic functions of our systems dynamics simulation model philosophy are described in the causal loop diagram (CLD) shown in Figure 3. The figure shows that with increased population, the consumption of metals increases, which in turn increases the production. Emissions and waste generated from both the production and consumption of the metals lead to environmental degradation. Increased environmental degradation increases public and governmental concerns and forces society to take necessary policy actions. These actions are shown in the CLD with numbers from 1 to 4 (blue diamonds) in Figure 3. Increasing consumption and population are the two major factors for an increasing demand for metals in the world. An increase in population drives consumption, depleting markets, increasing prices and increasing supply from production to market. This allows for continued consumption augmentation as well as increased resource use. Increased resource use rate and associated waste generation leads to environmental degeneration.

Environmental degradation and declining resources have an effect on political and public awareness. This leads to the development of four different policy options. During the early 1950's, end of pipe solutions (blue diamond no. 1 in Figure 3) were used as a first response to increased concerns over environmental degradation. Instead of draining out wastewater from industrial process to rivers, we built wastewater treatment plants; or instead of emitting hazardous waste gasses into the atmosphere, we installed treatment units in such processes. During the early 90's we realised the economic value of natural resources and waste, and introduced cleaner production and pollution prevention practices (diamond no. 2) to increase the efficiency in the production processes, and thus to decrease the use of raw materials (natural resources), the waste generated and gasses emitted to the atmosphere. In the last decade, we have concentrated on sustainable consumption and production behaviour (diamond no. 3) and begun to question how we can make changes in our life style (and quality) to decrease the demand for goods and food, and consume less, which may in turn eventually decrease the environmental degradation. As a part of sustainable production policy, recycling represents a way to increase metals in the societal material cycle without depleting resources. However, as can be seen from the CLD in Figure 3 - if we trace back the main root cause for today's increasing environmental degradation, it is embedded in the increase in the world's population. We certainly need to introduce sustainable population policies (diamond no. 4) (especially in the developing countries), together with sustainable consumption and production policies (diamond no. 3) (mainly in developed countries) in

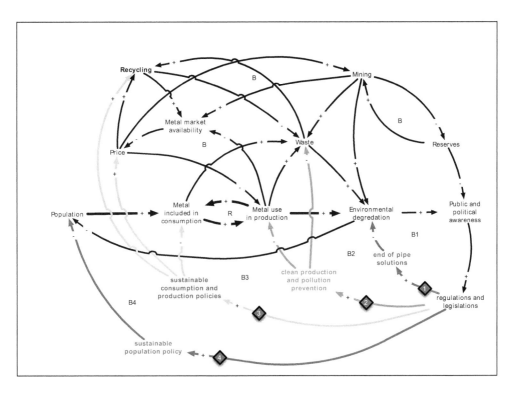

Fig. 3. Sustainability of resource use has moved over many system levels from end-of-pipe (fighting pollution) to root cause (population numbers and their behaviour). Attention has over time moved from end-of-pipe solutions (1) to more focus on clean production (2) recycling, slimmer consumption patterns and sustainable production (3). Ultimately the world must also address the consumption volume as a function of per capita use as well as the number of consumers, directly proportional to the size of the global population (4).

order to decrease over population (demand) and unnecessary wasting (supply). A long-term sustainability policy for the world population will thus be needed, as a part of the total flux outrunning planetary capacities. In this context, Figure 3 presents the problem and displays the different solutions tried so far. Unsustainability in this context arises from: (a) End of pipe pollution output from the system; (b) From unsustainable production or resource use in suboptimal products; (c) From excessive volume consumption of resources; (d) and - From consumption in excess of the carrying capacity of the Earth. In order to address the problems, four different approaches were tested in this study. Potentially, it may be that a global population contraction during the next century must be planned for (Malthus, 1798, Pearson & Harper, 1945, Osborn, 1948, Ehrlich, 1968, Meadows et al., 1972, 1992, 2004, Brown, 2009a,b, Ehrlich & Ehrlich, 1990a,b, 2006, 2009, Bahn & Flenley, 1992, Ehrlich et al., 1992, Daily & Ehrlich, 1992, Brown & Kane, 1994, Daily et al., 1994, Evans, 1998).

The carrying capacity of the world for population has been estimated many times, but with disparate results, primarily because of differences in fundamental assumptions (Cohen, 1995) concerning the following aspects:

1. Energy:
   a. Finite fossil fuels (hydrocarbons, nuclear)
   b. Renewable energy (water, wind, sun, wood, dung)
2. Metals
   a. For infrastructure and tooling (Fe, Al, Co, Zn, Mn, Cr, Ni...)
3. Land
   a. Area suitable for cultivation
   b. Uncultivated land with sustainably harvestable resources
4. Technological food production capability
   a. Technical equipment and machinery
   b. Materials for simple civil construction (brick, wood)
5. Essential resources for sustainable harvest of biomass
   a. Essential trace elements (Co, Mo, Cu, B, Zn, S,....)
   b. Nitrogen
   c. Phosphorus
   d. Water
   e. Base cations (Ca, Mg, K)
6. Social resources
   a. Suitable workforce under sustainable local social conditions
   b. Social conditions conducive to sustained activity through a growing season (absence of warfare, functioning markets, adequate transportation, social accountability, reasonable degree of law and order, personal security sufficient for food storage).

Penck (1925), based on the work of von Liebig (von Liebig et al., 1841, Liebig 1843), defined the basic equation for the number of people that can be fed, the maximum population, called "Liebig's law": *Harvest is limited by the nutrient in least supply*:

$$\text{Sustainable population} =$$
$$\text{Total resource available annually / individual annual consumption [persons]} \qquad (3)$$

The equation is applied if the resource is renewable. If it is neither renewable nor substitutable, but constitutes a one-time heritage, then the annual sustainability estimate is:

$$\text{Total resource available annually} =$$
$$\text{Total resource volume / time to doomsday [ton per year]} \qquad (4)$$

The time to doomsday is estimated as the time to the end of our consideration, potentially the projected time of eclipse of human civilization (Gott, 1994, Leslie, 1998, Korb & Oliver, 1998, Sowers, 2002, Sober, 2003). We have previously discussed the content and meaning of a long-term time perspective, and we also made an assessment of the impacts of unsustainability of the phosphorus supply with respect to the global population (Ragnarsdottir et al., 2011, Sverdrup & Ragnarsdottir, 2011). There we arrived at the following definitions that we adopt here:

- **Long term sustainable** perspective is when the resource is managed in such a way that a glacial gap can be bridged. Glaciation causes denudation and access to fresh strategic element bearing rock. The average time between glaciations, the room for civilization to prosper, is about 10,000 years. After that geologically large events make conditions for civilizations so fundamentally changed that no standard rules apply. The first urban societies arose 10,000 years ago, the first states emerged 5,000 years ago.
- A **semi-sustainable** timeframe is a time equivalent to how long modern literate and democratic civilizations as we know them have persisted, more than 2,000 years. However, most societies unconsciously want to persist longer than that. The state of Denmark as an entity is about 2,000 years old.
- A **sustainability-oriented** timeframe is the historic time that is set for 1,000 - 2,000 years, the time of continuous historical records and age of the oldest surviving books.
- **Unsustainability** in the intermediate term is 200 - 1,000 years, the time perspective of many monuments and infrastructures that are well built. Unsustainability in the short term is considered to be 100-200 years. This timeframe is bordering on living memory, and often the age of a private house still lived in.
- **Urgent unsustainability** is adopted as the term for time horizons less than 100 years.

Of note is that individual consumption is not seen as the individual physiological requirement, but that it must include efficiencies from first extraction from the deposit until it reaches the individual consumer.

$$\text{Individual supply} = \text{Extracted amount per capita} \times \text{product of all efficiencies in the supply chain} \ [\text{kg per person}] \qquad (5)$$

The extraction steps may be many and the inefficiencies may quickly pile up. The carrying capacity of the Earth under total sustainability will be in sustainable population number (Ragnarsdottir et al., 2011, Sverdrup & Ragnarsdottir 2011):

$$\text{Sustainable population} = \min (\text{Sustainable population estimates (food production limitations (i)))} \ [\text{persons}] \quad (6)$$

According to Liebig´s principle estimates represent the different aspects that can limit growth (nitrogen, phosphorus, water, light, essential major and trace elements, soil substrate availability). In the short run, many of those can be overrun as long as the system can deplete the available resources. However, in the long run this is impossible, as it would violate mass balance laws. Many studies have considered these resources one by one, a few have done several, but none have done them all (See Table 1 and Table 2 for an overview of different estimates). We also use a simple equation for total consumption of a resource:

$$\text{Total consumption} = \text{Consumption per individual} \times \text{number of individuals} \ [\text{ton per year}] \qquad (7)$$

It is evident that we can take down total consumption by reducing the amount each consumer uses, but also by reducing the number of consumers, and both. It is important to assess the effect of recycling. An integrated assessment over all essential components is needed in the long run, as the studies of Meadows et al. (2004). Thus, what we first estimate is the supply to society, as:

$$\text{Supply to society} = \text{Mining} \, / \, (1\text{-}R) \quad [\text{ton per year}] \qquad (8)$$

Where R is the degree of recycling on the flux from society. This is what is shown in Figure 4. In the calculations, we take the present mining rate, and use the present recycling degree, to estimate the present supply to society. Then, we calculate the new flow into society for other improved degrees of recycling:

$$\text{Time to scarcity} = \text{Reserves} \, / \, (\text{Supply to society} * (1\text{-}R_i)) \quad [\text{yrs}] \qquad (10)$$

where we have defined six scenarios (i), including:

1. Business-as-usual, no change in recycling from today's;
2. Improved habits in the market, at least 50% recycling or maintain what we have if it is higher than 50% recycling, and improving gold recycling to 95% recycling;
3. Improve all recycling to 90%, except gold to 96% recycling;
4. Improve all recycling to 95%, except gold, platinum, palladium and rhodium to 98% recycling;
5. Improve all recycling to 95%, except gold, platinum, palladium and rhodium to 98% recycling; assume same per capita use as in 4, but assume that population is reduced to 3 billion;
6. Improve all recycling to 95%, except gold, platinum, palladium and rhodium to 98% recycling; assume one half of present per capita use as in 4, but assume that population is reduced to 3 billion.

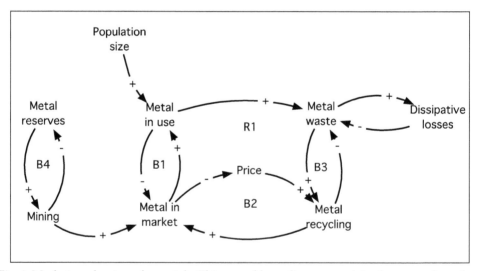

Fig. 4. Market mechanisms for metals. This causal loop diagram explains how supply and demand affect the price of a metal, and how the price feeds back on recycling and use. It can be seen from the diagram, that introduction of recycling creates a reinforcing loop, keeping material in the cycle. The enemy in the system is dissipative losses as they represent a destination for metals with no hope of return. The major driver for metal use is the population size.

Then we calculate the new net supply needed to maintain that societal supply at present level at improved recycling rates, and use that to find the new burn-off time. Figure 5 shows the flow diagram depicting what we explained above in the text, that recycling can maintain the same input to society, but decrease the input from finite resources through mining.

The real flow to society becomes amplified by recycling, because part of the outflow becomes returned to the inflow.

$$\text{Burn-off time for new recycling rate} = \text{reserves} / (\text{supply to society} \times (1-R_i)) [\text{yrs}] \quad (11)$$

Where R_i is the recycling of scenario i. In order to get the Hubbert's methods estimate of time to scarcity, conversion from burn-off based time to scarcity is (based on our results by plotting them):

$$\text{Hubbert's time to 10\% of peak production} = 1.7 * \text{Burn-off time} + 2 \quad (12)$$

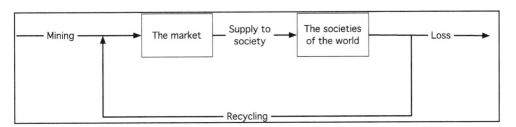

Fig. 5. The recycling effect. A flow diagram showing that recycling can maintain the same input to society, but decrease the input from finite resources through mining. The real flow to society becomes amplified by recycling, because part of the outflow becomes returned to the inflow.

Systems dynamics modelling of material cycles

A CLD neither demonstrates how strong each of the linkages is nor the functional form that the relationship takes. It allows us, however, to develop a better understanding of the overall system, to identify its main components as well as the cause-effect relations between these components and to trace back some main root causes of unsustainability in the system. In a fairly complex system, as the one presented here, there are substantial numbers of feedback loops, all of which result in a complex system behaviour. A model description of the CLD type does not contain all the details necessary for a full understanding of the model's behaviour, but it is possible to identify some overall reinforcing and balancing loops. We do not use neoclassical economic models in this study for analysing trends (Shafik, 1994, Turner, 2008, Goklany, 2009) as these methods do neither use mass and energy balances nor obey thermodynamic principles, thus yielding naïve results of limited value. It is important to state that neither econometricists nor economists can choose whether they believe or do not believe in thermodynamics. Time has come where we stop to be polite and tolerate their ignorance. There is no escape from mass balance, and not internalizing this mass balance has become damaging to society, "….its validity is absolute, leaving those that disbelieve in line for complete humiliation" as Sir Arthur Eddington stated (Eddington, 1928, see Hougen et al., 1959 for further elaboration on the consequence and danger of ignoring mass balances).

In this study, we use causal loop diagrams for finding each metal system connections, important feedbacks and system structures as a part of the generic systems dynamics procedure (Bertalanffy, 1968, Forrester, 1971, Meadows et al., 1972, 1992, 2004, Senge, 1990, Vennix et al., 1992, Sterman, 2000, Maani & Cavana, 2000, McGarvey & Hannon, 2004, Sverdrup & Svensson, 2002a,b, Cavana, 2004, Haraldsson & Sverdrup, 2004) and the learning loop (Haraldsson et al., 2002, 2007, Haraldsson 2007). The method used for constructing the system dynamics model follows a strict scheme, as well as deriving links by empirical-, experimental- and Delphi methods (Adler & Ziglio, 1996). The CLDs uniquely define the differential equations of the system, and together with a flowchart, help to build the dynamic models in the STELLA® modelling software.

Dynamic models are developed for gold, platinum, phosphorus, rare earth elements, lithium, uranium, thorium and oil. The models are then used to estimate time to scarcity, with and without recycling for different alternative scenarios. We define the time to scarcity as the time for the known reserves of high grade and low grade to have decreased to 10% of the original amounts.

The models are constructed into several modules:

1. The global population and consumption module;
2. The mining module with reserves and prospecting;
3. The consumption and market module, including a price mechanism;
4. The recycling module;
5. The social stress module.

Each model is formulated as a series of differential equations, arranged from mass balances for the resource R:

$$dR/dt = \text{inputs - outflows + produced - accumulation in the system} \quad [\text{ton per year}] \quad (13)$$

The following 10-12 different coupled reservoirs are considered in the models through coupled differential equations, based on mass balances for a number of stocks in the system:

a. Exhaustible resources (4-6 stocks):
    1. High grade deposits
    2. Hidden high grade deposits
    3. Low grade deposits
    4. Hidden low grade deposits
b. Population model (simplified global) (4 stocks) (0-20 yr, 21-44 yr, 45-65 yr, 65+ yr)
c. Society (2 stocks)
    5. Market stock
    6. Waste stock

For phosphorus, gold and platinum, we also used two additional physical reservoirs, known ultralow grade and hidden ultralow grade, as well as one for social trust in the phosphorous model. The material flow chart for gold including the recycling is shown in Figure 6. The causal loop diagram given in Figure 7 shows the causal chains used in our model on global gold supply. The CLD shows the basics of the market, but for simplicity omitting the derivates trade that also belongs to the model.

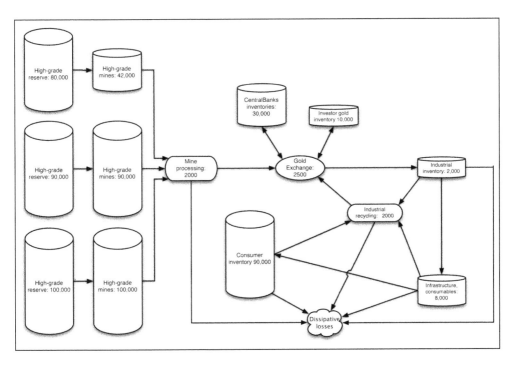

Fig. 6. Flow chart for world gold material fluxes. Actors trade through the market, trade takes place geographically dispersed, but linked through the price systems at the London and New York Metal exchanges. The numbers indicate approximate amounts of gold in metric tons in early 2009. The flowchart layout looks similar for most other metals.

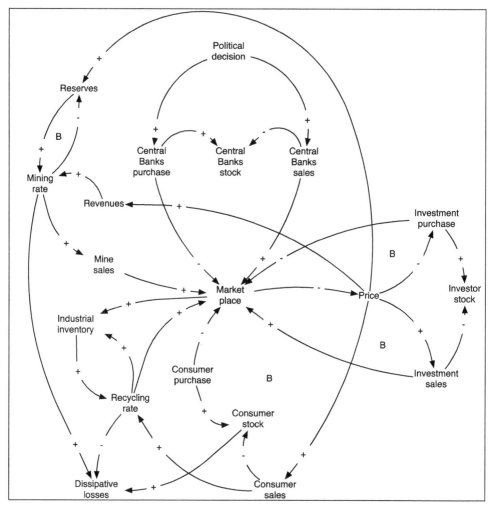

Fig. 7. The causal loop diagram underlying the models used in this chapter. The simplified CLD shows the basics of the market omitting the derivates trade that also belongs in the model.

3. Results

Material cycles and time horizons

Table 1 shows an overview of important metals and elemental resources for running society and human civilizations. The main sources and the main uses are listed, together with indications on the use of the reserves. Comprehensive life-cycle flow assessments are available for few substances. This is primarily a fact because many of the mining companies closely guard their data, making world reserve figures uncertain (USGS, 2008, Roskill Information Services, 2007a,b,c,d, 2008, 2009a,b, 2010a,b,c,d, 2011). However, some

Material	Source country	Main uses
		Bulk materials for society
Iron	India, Russia, Brazil, Germany, France, Sweden	Construction materials, machinery, vehicles, weaponry, household items, containers, building reinforcement (Forester, 1971, Meadows et al., 1972, 1992, 2005, Graedel et al., 2004). There is lots of aluminium in solid rock, but this is so tightly bound, that it is prohibitively expensive to take them out. Large losses are to corrosion and in household trash
Aluminium	Guinea, USA Australia, Canada, Russia, Norway	Aircraft, packaging, wiring, lightweight machinery, transport vehicles, electrical consumer durables, cars, rocket fuel, military equipment, soda-pop and beer cans, packaging, wiring (Forester, 1971, Meadows et al., 1972, 1992, 2005, Graedel et al., 2004, Roskill Information services, 2008, USGS, 2008, 2011)
Nickel	Canada, Russia	Steel alloys, galvanizing, tools, coins, LNG tanks, military equipment and weaponry (Forester 1971, Meadows et al., 1972, 1992, 2005, Graedel et al., 2004, International Nickel Study group 2008, USGS, 2008, 2011)
Manganese	South Africa, Brazil	Steel alloys, tools. (Forester, 1971, Meadows et al., 1972, 1992, 2005, Graedel et al., 2004, USGS, 2008, 2011)
Copper	Chile, Russia, Congo, Malawi, Zambia	Wiring, appliances, tubing, household items, major constituent of brass, coins (Forester, 1971, Meadows et al., 1972, 1992, 2005, International Copper Study group 2004, Graedel et al., 2004, USGS, 2008, 2011)
Zinc	China, USA, Austria	Galvanizing, batteries, brass, anti-corrosion, infrastructural items (Forester 1971, Meadows et al., 1972, 1992, 2005, International Lead and Zinc Study group 2003, Graedel et al., 2004, USGS, 2008, 2011)
		Strategic metals and elements
Gold	South Africa, Ghana, Mali, China, Russia, Canada	Investment, jewelry, gold plating, catalyst, coins, electronics contacts, circuit boards, semiconductor wiring, mobile phone antennas, dental materials (Forester, 1971, Meadows et al., 1972, 1992, 2005, Graedel et al., 2004, Heinberg 2007, Sverdrup et al., 2011a, USGS, 2008, 2011, The Gold Council website, unpublished industrial data)
Silver	Mexico, Bolivia, Chile	Silver plating, industrial conductors, wiring, jewlery, financial placement, coins, hollowware, cutlery (Forester, 1971, Meadows et al., 1972, 1992, 2005, Cross 2000, Graedel et al., 2004, Heinberg 2007, Sverdrup et al., 2011, USGS, 2008, 2011, unpublished industrial data)
Wolfram	China, South Africa	Lamp filaments, high temperature applications, weaponry, cutting materials (Forester 1971, Meadows et al., 1972, 1992, 2005, Graedel et al., 2004, Roskill Information Services 2007b, Ragnarsdottir, 2008, USGS, 2008, 2011).
Helium	USA, Algeria, Qatar, Russia,	Research, superconductors, ballons, protective gas (Forester 1971, Meadows et al., 1972, 1992, 2005, Graedel et al., 2004, USGS, 2008, 2011)
Indium	Canada, China, USA,	LCDs, new generation of solar cells, microprocessors (Ragnarsdottir 2008, USGS, 2008, 2011, Roskill Information Services 2010b). Byproduct of Zn.
Tin	China, Brazil, Malaysia	Cans, solder, paint. (Forester 1971, Meadows et al., 1972, 1992, 2005, Graedel et al., 2004, Ragnarsdottir, 2008, USGS, 2008)
Tantalum	Australia, USA, Brazil, Congo	Mobile phones, camera lenses, DVD players, computers, super-alloys, piping, minor component of brass (Ragnarsdottir, 2008, USGS, 2008, 2011, Roskill Information Services 2009a)
Titanium	South Africa, Australia, Norway, Russia, USA	Light and strong metal, and at ambient conditions, it is very resistant to corrosion. It is produced for specialized technologies and as titanium oxide as a white colour pigment. The production is difficult and energy demanding. There is a lot of titanium on Earth, however, the mineral deposits where extraction can be undertaken, are limited. (Roskill Information Services 2010a)
Niobium	Australia, USA, Brazil, Congo	Specialty alloys for high temperatures, engine turbines (Graedel et al., 2004, USGS, 2008, Roskill Information Services 2009b)
Lead	China, Australia, USA	Pipes, batteries, soldering, alloys, weaponry (Meadows et al., 1972, 1992, 2005, International Lead and Zinc Study group 2003, Graedel et al., 2004, USGS, 2008)
Lithium	Bolivia, Chile, Tibet	Batteries, medicine, nuclear bombs (Meridian International Research 2008, USGS, 2002, USGS, 2008, 2011)
Rare earths	Canada, China, South Africa, Australia, Norway, Russia, Brazil	Compact batteries, LED's, miniature magnets, catalysts, new technologies, optics, specialty alloys, steel, lasers (Graedel et al., 2004, Roskill Information Services 2007a, Ragnarsdottir 2008, USGS, 2008, 2011, unpublished industrial data). Rare earths reserves are often associated with co-deposition of phosphates, and part of these reserves represents back-up reserves for phosphate
Yttrium	Canada, China, South Africa, Australia, Norway, Russia, Brazil	Flat screens and cathode ray tube displays and in LEDs. Production of electrodes, electrolytes, electronic filters, lasers, pinhead capacitors and superconductors; various medical applications; as traces in materials as grain size modifier. In superconductors. Yttrium garnets for optic filters, glass for telescopes (Graedel et al., 2004, Ragnarsdottir, 2008, USGS, 2008, 2011, Roskill Information Services 2009, unpublished industrial data)
Platinum	South Africa, Russia, Canada	Jewelry, fertilizer catalyst, car catalyst, car fuel cells, dentistry, implants (Graedel et al., 2004, Johnson Matthey's Platinum Review, 2008, Ragnarsdottir 2008, USGS, 2008, 2011, Sverdrup et al., 2011a,b, Sverdrup & Pedersen, 2009, unpublished industrial data)
Palladium	South Africa, Russia, Canada. Depends on Pt production	Jewelry, fertilizer catalyst, car catalyst, car fuel cells, dentistry, implants, laboratory gear (Graedel et al., 2004, Johnson Mathey's Platinum Review, 2008, Ragnarsdottir, 2008, USGS, 2008, 2011, Sverdrup et al., 2011a,b, unpublished industrial data)
Rhodium	South Africa, Russia, Canada,	Fertilizer catalyst, car catalyst, car fuel cells (Graedel et al., 2004, Johnson Mathey's Platinum Review, 2008, Ragnarsdottir, 2008, USGS, 2008, 2011, Sverdrup & Pedersen, 2009, Sverdrup et al., 2011a,b, unpublished industrial data). By product of Pt production
Cobalt	Congo, Zambia, Australia, Canada, Russia	Used for magnetic alloys, wear-resistant and high-strength alloys, and as a catalyst in several chemical processes. Blue colour in glass, ceramics, ink and paints (Roskill Information Services 2007c)
Germanium	China, Russia	Electronics, LED, laser, superconductors (USGS, 2008, 2011). Mostly by-product of Zn ores
Gallium	China	Electronics, LED (USGS, 2008, 2011, Roskill Information Services 2011)
Arsenic	China, Chile, Peru, Morocco	Antifouling agent, poison, wood preservative agent, electronics, LED, laser, superconductors, medicine, in copper and zinc-based alloys. We are not dependent on this toxic substance. The use is declining (USGS, 2008, 2011)
Tellurium	Canada, USA, Peru, Japan	Used in alloys, infrared sensitive semiconductors, ceramics, specialty glasses, and photovoltaic solar panels (USGS, 2008, 2011)
Antimony	United States, China	Solder, semiconductors, gunshot substitute (Roskill Information Services 2007d, USGS, 2008, 2011). Alloys with lead, as a flame retardant and as a catalyst in organic chemistry. New uses are in transistors and diodes
Selenium	Japan, Belgium, Russia, Chile	Used in photovoltaic cells and photoconductive action, where the electrical resistance decreases with increased illumination. Exposure meters for photographic use, as well as solar cells and rectifiers, p-type semiconductor and electronic and solid-state applications. Used in photocopying. Used by the glass industry to decolourise glass and to make ruby coloured glasses and enamels. Photographic toner, additive for stainless steel. Mostly by-product of copper ores (USGS, 2008, 2011). http://www.webelements/selenium/uses.html. Selenium is an essential nutrient – necessary for a strong immune system
Rhenium	Chile, USA, Kazakstan, Peru	Catalysts, specialty alloys for jet engines (Roskill Information Services 2010c)
		Sources of energy
Oil and Gas	Saudi-Arabia, Emirates, Iraq, Kuwait, Iran, Canada, Russia, Venezuela, Mexico, Indonesia	Energy for transportation, only operational available propellant for aircraft, propellant for ships, propellant for cars, for industry, raw material for plastics, heat source for cement, for steel industry, for domestic heating (Pogue & Hill 1956, Meadows et al., 1972, 1992, 2005, Hubbert, 1982, Arleklett, 2003, 2007, Greene et al., 2003, Hirsch, 2005, Energy Information Administration, 2007, Heinberg, 2007, Strahan, 2007, 2008, Zittel & Schindler, 2007, BP, 2008, USGS, 2008, 2011, unpublished industrial data)
Coal	All continents	Steel and metal production, cement, electricity, heating (Pogue & Hill 1956, Strahan, 2007, 2008, Zittel & Schindler 2007, USGS, 2008, 2011, unpublished industrial data)
Uranium	Russia, Africa, China, Canada	Conventional nuclear energy, atomic weaponry, potentially in breeder reactors (Francois et al., 2004, USGS, 2008, 011)
Thorium	Canada, China, Australia, Norway, India, South Africa	Nuclear energy, potentially in breeder reactors, closed cycle nuclear energy (Jayaram, 1985, Kasten, 1998, USGS, 2008, 2011, unpublished industrial data)
		Essential for human life support
Phosphorus	Morocco, South Africa, China, Russia, Australia	Fertilizer for food production, pesticides (Meadows et al., 1972, 1992, 2005, Fillipelli, 2002, 2008, Oelkers & Valsami-Jones, 2008, USGS, 2008, 2011, Brown, 2009, Ragnarsdottir et al., 2011, Sverdrup & Ragnarsdottir, 2011)

Table 1. Overview of important metals and elemental resources for running society and human civilizations.

material-flow studies detailing a series of life stages are available. Mining and processing, fabrication, use and end-of-life data are documented for copper, zinc and lead (Graedel et al., 2002, Gordon et al., 2003, 2004, 2006, Spatari et al., 2005, Mao et al., 2008), with less-detailed information available for platinum (Råde, 2001). Information is available in special industrial publications (Johnson Matthey's Platinum Review, annual journals 1980-2011 were consulted) for tin, silver and nickel (Gordon et al., 2003, 2006) and indium and gallium (Cohen 2007, Ragnarsdottir, 2008, Roskill 2010b, 2011). Data for gold is available in official and unofficial statistics, and can mostly be found through web searches.

In Table 2 we show some of the outputs of our calculations. They show estimated burn-off times, time to scarcity by using the adapted empirical approach of Hubbert (1966, 1972, 1982), and using integrated systems dynamics modelling for metals and elements of different classes. In Table 2, "#" represents estimates using dynamic simulation models built by the authors, where as "*" represents earlier dynamic model assessments by Meadows et al. (1972, 1992, 2004). The reserve values given represent the sum of estimates for high-grade and low-grade ores. Platinum, palladium and rhodium are being used notably in catalytic converters that make burning of fossil fuels more efficient, and to catalyse a plethora of reactions in the chemical, petrochemical and pharmaceutical industries, and are at the core of ceramic fuel cells. Platinum is key to the production of fertilizer for food production and therefore a part of food security. Platinum group metals have been estimated to be of the order of maximum 80,000 ton, of which about 17,000-25,000 ton are effectively available after taking into account the efficiencies of the different mining, milling and smelting processes (Råde, 2001). With petroleum becoming more scarce and expensive, fuel cells using platinum group elements as catalysts are seen as future sources of alternative energy. But the reserves of these elements will not fuel the world's cars long into this century simply because there is not enough resource to power the cars.

For platinum and palladium, there are dedicated platinum group metal mines (South Africa), but about 30% of the production is a by-product of the nickel production (Russia, Canada). Rhodium production is entirely dependent on the production of platinum and palladium, and when they stop, so does rhodium (Johnson Matthey Platinum Review, 2008). In the systems dynamic modelling output diagrams as shown in Figure 7, we can see dynamic simulations undertaken for platinum, with past and predicted future reserves in geological formations to the left and mining rate to the right. Platinum shows peak behaviour, similar to what many other materials do. Platinum is lost in several ways. Only about 20-25% of the platinum metal used for catalytic converters for cars is recycled every year. This it out of a flux of 40 ton per year, where the global production is 230 ton per year. 40-50 ton is used as catalyst in fertilizer plants; of this 10-20% is lost diffusively into the produced product (4-8 ton per year). Overall global recovery is perhaps in the range 65-75%, which needs to be improved.

Gold has a very special place amongst the metals, as it is the first metal ever used by humans. It is in every day life easy to work with and nearly indestructible. It is traded as money, but has many technical applications, the most common being decorative plating, and plating for protection against corrosion, in jewelry and electronics. It is also special in the sense that the production is well known, it has passed the peak, and we now have more above ground than in geological formations. Gold is very valuable, has always been so, and almost nothing is lost aside from small dissipative losses. It's first and most important use was as money and in prestige jewelry; this remains so to the present. In times of inflation and governmental unaccountability with issuing fiat money, gold is the resort taken by a

Element	Burn-off time, years	Hubbert-time to 10%, years	Dynamic model, years	2008 Available deposits, ton	Estimates 2008 mining rate ton/year	Present Recycling %
Bulk materials for societal infrastructures						
Iron	79	176	200*	150,000,000,000	1,900,000,000	20
Aluminium	132	286	300*	25,000,000,000	190,000,000	30
Nickel	42	95	300*	67,000,000	1,600,000	50
Copper	31	71	120*	490,000,000	15,600,000	50
Zinc	20	38	40*	180,000,000	10,500,00	10
Manganese	19	78	50*	300,000,000	8,800,000	20
Strategic materials for technology						
Indium (Zn-dependent)	25	43		11,000	580	0
Lithium	25	75	330#	4,900,000	200,000	0
Rare earths (Ce, La, Nd, Pr, Sa, Eu, Gd, Tm, Tb, Lu, Tb, Er)	455	400-900	1,090#	100,000,000	120,000-220,000	5
Yttrium (REE dependent)	61			540,000	8,900	20
Hafnium (Zr-dependent)	6200	132		310,000	50	80
Zirconium	67	152		60,000,000	900,000	20
Tin	20	45	60*	6,100,000	300,000	26
Molybdenum	48	120		8,600,000	180,000	25
Rhenium (Mo-dependent)	50	110		2,500	50	75
Lead	23	51	45*	79,000,000	3,500,000	60
Wolfram	32	74	90*	2,900,000	90,000	20
Cobalt	113	255		7,000,000	62,000	20
Tantalum	171	395		240,000	1,400	20
Niobium (Mo-dependent)	45	95		2,700,000	60,000	20
Helium	9	19		7,700,000	882,000	0
Chromium	86	100		18,000,000	210,000	25
Gallium	500			100,000	200	30
Arsenic	31	55		1,700,000	55,000	0
Germanium	100	210		4,000	40	30-35
Titanium	400			600,000,000	1,500,000	50
Tellurium	387			58,000	150	50
Antimony	25	50		5,000,000	200,000	30
Selenium	208			250,000	1,200	0
Precious metals						
Gold	48	37	75#	100,000	2,100	98
Silver	14	44		400,000	28,000	70
Platinum (Ni-dependent)	73	163	50-150#	16,000	220	70
Palladium (Pt-dependent)	61	134		14,000	230	70
Rhodium (Pt-dependent)	44	108	50-150#	1,100	25	60
Fossil energy resources						
Oil	44	100	99#,*	164,000,000,000	3,700,000,000	0
Coal	78	174	220#,*	470,000,000,000	6,000,000,000	0
Natural gas	64	143	100*	164,000,000,000	2,600,000,000	0
Uranium	61	142	180*	3,900,000	64,000	0
Thorium	187	140-470	335#,*	6-12,000,000	64,000	0
Planetary life support essential element						
Phosphorus	80	95-285	230-330#	6-12,000,000,000	145,000,000	10-20

Table 2. Estimated burn-off times (years), time to scarcity using the empirical approach of Hubbert (1966, 1982), and using integrated systems dynamics modelling for metals and elements of different classes. # represents estimates using models built by authors, * represents dynamic model assessments by Meadows et al. (1972, 1992, 2004) using system dynamics. Available deposit values represent the sum of estimates for high grade and low-grade ores.

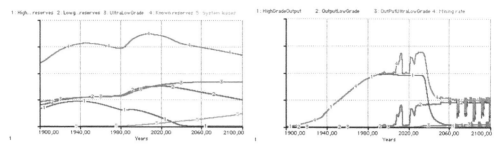

Fig. 8. Peak platinum. Systems dynamics modelling output diagrams showing past and predicted future platinum reserves in geological formations (to the left) and mining rate (to the right).

majority of the world's population. Such practice becomes more prevalent in times of uncertainty. Gold world production peaked in 2005 at 2,500 ton per year, and is expected to decline further from now (now 2,100 ton per year). Overall the recycling degree of gold metal in industrial circulation is better than 92%. In modern times, significant amounts are being lost through gold plated objects and in consumer electronics dumped in landfills and burned in incinerators. Total losses are estimated at 12,000 ton over 5,000 years out of a total mined volume of 160,000 ton (7.5%). A complicating affair with gold is that much has been sold forward as paper gold to investors; this implies that the ownership is sold, but the actual physical metal has not been delivered. This is undertaken by a number of investment and hedge-fund banks that have no significant reserves of physical gold, and thus the gold they sell probably does not exist. They simply assume that they can buy it "somewhere" whenever the need should arise. For a physically limited commodity as gold, that may not be the case at all times. This situation of uncovered gold forward positions will continue to put pressure on the gold price and thus physical supply and demand for decades to come. In the systems dynamic modelling output diagrams in Figure 9, we can see dynamic simulations undertaken for gold, with past and predicted future reserves in geological formations to the left and mining rate to the right.

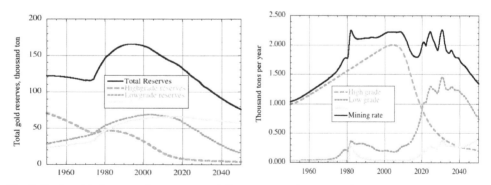

Fig. 9. Peak gold. Systems dynamics modelling output diagrams showing past and predicted future gold reserves in geological formations (to the left) and mining rate (to the right).

Figure 10 shows simulated past and future market price for platinum and gold. The metals will never become unavailable when the resource reserves dwindle, but they will become more expensive as the market availability goes down. The model was validated against observed price as well as on central banks inventory statistics in the past with good success.

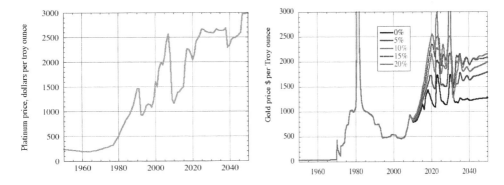

Fig. 10. The price-impact of metal scarcity. System dynamics modelling output diagrams showing past and future market price for platinum (to the left) and gold (to the right). The different % represents at which average rate uncovered forward and short positions are eliminated.

Silver is an important industrial metal, because of its good ductility and conductivity to electricity and heat. As with gold, significant amounts of silver have been speculatively sold forward by banks, and it is uncertain how this is to be found for physical delivery in the future when the contracts expire (OCC, 2009). Silver is lost to diffusive losses, in electronics and silver plating. Cutlery and coinage is mostly recycled. There are considerable amounts of silver stored in private homes all over the world, and a small calculation assuming that in the western world every home has 200 g silver (4 spoons, one in 20 has a full, but simple silverware set) and in the rest of the world there is 15 g silver per person (one teaspoon per household, one in 200 has a full, but simple silverware set), an estimated 400,000 tons is stored in private homes, or about 14 annual global productions.

Rare earths are always mined together as a mix; they are very difficult to separate; with them comes often Y, Sc, Ni and Ta. Figure 11 shows model outputs from simulations for the peak behaviour of rare earth metals without (left) and with (right) recycling. It can be seen that with recycling, we can keep an equal amount available in the market with less mining and that the reserves last significantly longer. With recycling, the availability of the rare earths can be brought to last up to 50 times longer, depending on the degree of recycling.

Lithium is extracted from the salt-beds of certain dried out lakes at some few locations in the world. It has great potential for making light-weight batteries for computers, mobile phones, but possibly cars are also being contemplated. However, the resource base is very small, and mass production of accumulators for cars would quickly finish it off. At present, there is no recycling, and all lithium used at present is lost.

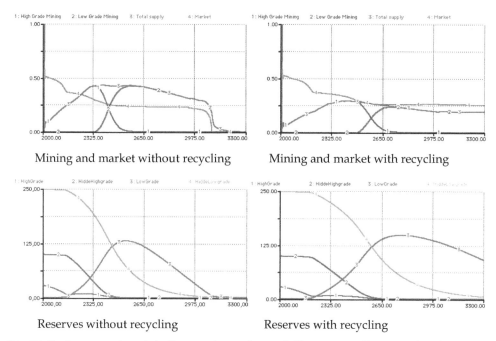

Fig. 11. Peak rare earth metals. System dynamics modelling output diagrams showing rare earth metals without (to the left) and with (to the right) recycling. Diagrams on top show the supply and mining rates, where as diagrams at the bottom show reserves over time.

Recycling and replacing

In view of these findings, it is essential that we use metals like copper and zinc, platinum group elements and tantalum as well as all other scarce resources in a sustainable manner. More efficient metal use through better product design and higher durability will also help. But there are limits to recycling; zinc, for example, is used in low concentrations for galvanizing, and is difficult to recycle once it is dissipated across metallic surfaces. Where possible, therefore, scarce resources need to be replaced by more common and easily accessible ones such as aluminium or silicon. A few such technologies are already being investigated. During the past century, industrial production around the world has increased 40-fold. Virgin stocks of several metals seem inadequate to provide the modern "developed world" quality of life for all people on Earth under contemporary technology (Gordon et al., 2006). Of course, there will always be lower-grade metal to exploit, and we can also mine at deeper levels, leading some Earth scientists to reject this more pessimistic view (Williams, 2008). But the price may well prove prohibitive in practice owing to the cost of energy.

Pin-head capacitors in mobile phones that are currently made out of yttrium, tantalum or hafnium, could instead use aluminium[2]. Germanium and indium transistors can possibly be made from silicon or silicon carbide (Juang et al., 2008) or carbon nanotubes (Tans et al., 1998). Solar cells made with silicon polyvinyl chlorides are twice as efficient as thin films solar cells made from the rare metals indium and gallium. For phosphorus there is neither a

---

[2] http://en.wikipedia.org/wiki/Electrolytic_capacitor

substitute nor any alternative. Economists and economic theorists frequently advocate that technology and science will always develop substitutes, but for phosphorus, we know as a fact that there is none at all because nothing can replace this essential element for life, regardless of what the price might be.

Phosphorus is unique and is demanded for maintaining life in all living organisms in rigidly set proportions. Once elements and metals are dissipatively lost, the energy cost of retrieving them will normally be prohibitive, and if the need is large, the energy needed will probably not be available (Ragnarsdottir, 2007, Fillipelli, 2008, Ragnarsdottir et al., 2011, Sverdrup & Ragnarsdottir, 2011). Extracting phosphorus from the seawater would be energetically unsolvable on a global supply scale. Systems dynamics outputs are shown in Figure 12. The figures show that mining rates from different sources in the longer perspective, show peak behaviour, how the production from high grade, then low grade and last the ultralow grade reserves, peaks (Ragnarsdottir & Sverdrup, 2011).

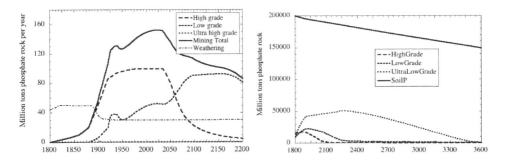

Fig. 12. Peak phosphorous. System dynamics modelling output diagrams showing how the mining rates from different sources peak (to the left) and how the production from high, low and ultralow grade reserves peak (to the right) in the longer perspective.

Job market is affected

It is not only price rises of metals on the world market that give an indication of scarcity: the Denver Post[3] reported in March 2008 that in mining companies, geologists were receiving higher average starting salaries than those with MBA degrees. Similar opportunities apply to skilled oil prospecting geologists and reservoir engineers. At present, these are put to work to make the system even more unsustainable, by finding more of the last hidden reserves, so that they can be finished off as well. However, as times go by and more and more resources can be shown to pass their production peak, the message will have to sink in that recycling is the new kind of mining. The new type of prospecting geologist will have to learn how to mine societal stocks to serve sustainability. Materials engineers and advanced metallurgists have also become rare (difficult subject to master well, few take on this

---

[3] http://www.denverpost.com/business/ci_8595906

education), and salaries for these are rising. In the near future more systems analysis and sustainability planners as well as material scientists will be needed across the world.

Convergence and contraction needs

From the estimates, using burn-off years, the Hubbert-curve estimate or systems dynamic model runs, we see that we are now challenging the planetary boundaries for supplying most of the materials humans consider necessary for serving civilization. Even metals like iron or aluminium may eventually run out in a few decades. It is of little help to point out that there exists an abundance of these in granite or in the deep crust, as long as there are no reasonable methods or technologies available for winning them. It is also evident from the systems analysis and the dynamic runs, that the market alone cannot cause the use of scarce resources to become sustainable. This is because the market is opportunistic in its function and nature, has no memory, or future vision. The rise in price when a resource becomes scarce will cause recycling to increase, but this occurs when too much of the resource has been consumed without significant recycling, and thus allows a large part of it to have become wasted. In addition to a well functioning market, strict governance and policies are needed. Governance must see to that the free market has a well-regulated arena to operate in as well as enforcing that those game rules are obeyed. This is needed to govern responsible use of resources, including starting recycling before the resource becomes scarce. Governments must make policies that look forward with responsibility for generations to come and for the preservation of society.

It is a widely spread misconception that a free market is a market without rules and regulations, with no interference from government; the opposite is the case (Smith, 1776, Friedman, 1962, Forrester, 1971, Friedman & Friedman, 1980, Sterman, 2000, Klein, 2007, Lövin, 2007, Sachs, 2008). The demand for free economy to mean no rules is nonsensical, even if it is cherished by certain political ideologies and taught at many business schools. Games without rules soon deteriorate to anarchy or rule of the strongest, and markets without rules quickly become something that has nothing to do with free markets (Friedman, 1962, Klein, 2007, Sachs, 2008, Jackson, 2009). Sustainability constraints are among some of the most important additions to free market economies if they are to be made long term stable and long term sustainable. The modern world depends on the markets systems for distribution and redistribution of goods, services and wealth, thus functioning markets are an integral part of a sustainable world.

In Table 3, we show the estimates of the maintenance supply from external sources to keep the internal society flux as in 2008, but applying a strict recycling principle. Here we have chosen scenario 4, applying 95% recycling for all elements, except precious metals that we set at 98% recycling. This is a very strict scheme, but it was chosen for the sake of example, to show how large the impact of recycling can be. Obviously, recycling is key factor in going towards metals sustainability. The required external supply can be brought down significantly, once recycling is made efficient. This, together with a contraction of the global population may bring us into the sustainability realm.

In Table 4 we show the outputs on Hubbert-estimates for time to scarcity. The scenarios are as follows; (1) Business-as-usual, no change in recycling from today; (2) At least 50% recycling or maintain what we have if it is higher than 50% recycling, and improving gold recycling to 95%; (3) Improve all recycling to 90%, except gold is improved to 96% recycling; (4) Improve all recycling to 95%, except gold, platinum, palladium and rhodium is

Element	Estimates 2008 mining rate (ton/year)	Recycling now (%)	Real supply to society 2008 (ton/year)	Scenario 4, recycling (%)	Required external supply to keep the real society flux, using scenario 4 recycling (ton/year)
Bulk materials for societal infrastructures					
Iron	1,900,000,000	20	2,300,000,000	95	120,000,000
Aluminium	190,000,000	30	270,000,000	95	13,570,000
Nickel	1,600,000	50	3,200,000	95	160,000
Copper	15,600,000	50	31,000,000	95	1,560,000
Zinc	10,500,000	10	13,100,000	95	656,000
Manganese	8,800,000	20	9,700,000	95	489,000
Strategic materials for technology					
Indium (Zn-dependent)	580	0	580	95	29
Lithium	200,000	0	200,000	95	10,000
Rare earths	220,000	5	232,000	95	11,600
Yttrium (REE dependent)	8,900	20	9,500	95	445
Hafnium (Zr-dependent)	50	80	250	95	13
Zirconium	900,000	20	1,150,000	95	56,250
Tin	300,000	26	405,000	95	20,270
Molybdenum	180,000	25	240,000	95	12,000
Rhenium (Mo-dependent)	50	75	200	95	10
Lead	3,500,000	60	8,700,000	95	437,000
Wolfram	90,000	20	112,500	95	5,625
Cobalt	62,000	20	77,500	95	3,875
Tantalum	1,400	20	1,750	95	88
Niobium (Mo-dependent)	60,000	20	75,000	95	3,750
Helium	882,000	0	882,000	95	44,100
Chromium	210,000	25	280,000	95	14,000
Gallium	200	30	285	95	14
Arsenic	55,000	0	55,000	95	2,749
Germanium	40	30	57	95	3
Titanium	1,500,000	50	3,000,000	95	140,000
Tellurium	150	50	300	95	15
Antimony	200,000	30	282,000	95	14,200
Selenium	1,200	0	1,200	95	24
Precious metals					
Gold	2,100	80	14,000	98	280
Silver	28,000	70	93,300	98	1,867
Platinum (Ni-dependent)	220	70	733	98	15
Palladium (Pt-dependent)	230	70	766	98	15
Rhodium (Pt-dependent)	25	60	83	98	2
Planetary life support essential element					
Phosphorus	145,000,000	20	187,500,000	95	50,000,000

Table 3. Estimates of the maintenance supply from external sources to keep the internal society flux as in 2008, but applying a strict recycling principle. The required external supply can be brought down significantly, once recycling is made efficient. The maintenance needs after this much recycling amounts to approximately 1/10 of the 2008 mining. This, together with a contraction of the global population may brings us into the sustainability realm.

Element	(1); BAU	(2); 50%	(3); 90%	(4); 95%	(5); 95%+3bn	(6); 95%+3bn+ ½
Bulk materials for societal infrastructures						
Iron	158	254	1,285	2,574	6,007	12,014
Aluminium	132	372	1,876	3,756	8,764	17,528
Nickel	82	82	424	851	1,986	3,972
Copper	61	61	317	638	1,488	2,975
Zinc	38	72	372	748	1,745	3,490
Manganese	35	58	306	616	1,437	2,874
Strategic materials for technology						
Indium (Zn-dependent)	35	74	385	771	1,798	3,597
Lithium	47	97	496	997	2,325	3,597
Rare earths	924	1,759	8,809	17,622	41,117	82,235
Yttrium (REE dependent)	120	616	1,235	2473	5,770	11,541
Hafnium (Zr-dependent)	12,649	12,649	25,303	50,609	118,087	236,174
Zirconium	133	214	1,085	2,173	5,071	10,142
Tin	38	58	304	611	1,425	2,850
Molybdenum	94	289	728	1,459	3,405	6,809
Rhenium	99	99	252	507	1,183	2,365
Lead	43	43	181	365	852	1,703
Wolfram	62	102	523	1,049	2,447	4,894
Cobalt	227	365	1,840	3,683	8,594	17,188
Tantalum	346	556	2,795	5,594	13,053	26,106
Niobium (Mo-dependent)	88	143	731	1,466	3,420	6,841
Helium	14	32	175	353	823	1,647
Chromium	175	262	1,310	2,600	6,100	12,200
Gallium	1,017	1,425	7,139	14,282	33,325	66,650
Arsenic	60	123	627	1,258	2,936	5,872
Germanium	201	282	1,425	2,854	6,659	13,317
Titanium	813	813	4,078	8,160	19,039	38,079
Tellurium	784	784	3,942	7,888	18,405	36,809
Antimony	48	68	354	711	1,658	3,317
Selenium	422	8,500	10,600	21,200	49,600	99,200
Precious metals						
Gold	94	94	142	725	1,693	3,385
Silver	26	26	84	434	1,012	2,024
Platinum	145	145	442	2,223	5,187	10,400
Palladium	121	121	369	1,860	4,340	8,679
Rhodium	86	86	266	1,343	3,135	6,269
Fossil energy resources						
Oil and gas	100	-	-	-	330	660
Coal	174	-	-	-	574	1,150
Uranium	121	240	1,215	12,184	28,400	56,900
Thorium	379	747	3,746	37,500	87,500	175,000
Planetary life support essential element						
Phosphorus	160	258	1,303	6,527	15,200	30,460
Colour legend						
TTS range, years	0-100	100-200	200-1,000	1,000-2,000	2,000-10,000	>10,000
Colour code						

Table 4. Outputs on Hubbert-estimates for time to scarcity. (1) Business-as-usual, no change in recycling from today's values, (2) Improved habits in the market, at least 50% recycling or maintain what we have if it is higher than 50% recycling, and improving gold recycling to 95% (3) Improve all recycling to 90% except gold to 96%. (4) Improve all recycling to 95%, except gold, platinum, palladium and rhodium to 98%. (5) Improve all recycling to 95%, except gold, platinum, palladium and rhodium 98%, assume same per capita use as in 4, but assume that population is reduced to 3 billion. (6) Improve all recycling to 95% except gold, platinum, palladium and rhodium to 98%, assume one half of present per capita use as in 4, but assume that population is reduced to 3 billion.

improved to 98% recycling; (5) Improve all recycling to 95%, except gold, platinum, palladium and rhodium to 98% recycling, assume same per capita use as in 4, but assume that population is reduced to 3 billion; (6) Improve all recycling to 95%, except gold, platinum, palladium and rhodium to 98% recycling, assume one half of the present per capita consumption with respect to scenario 4, but assume that the global population is reduced to 3 billion (Brown, 2009, Sverdrup & Ragnarsdottir, 2011). The colour code in the table shows the time to scarcity years. Degree of sustainability can be assessed as 'very unsustainable' for red and orange colours, 'unsustainable' for light orange, 'problematic for future generations' for yellow, 'approaching sustainability' for light green and 'fully sustainable' for green colours.

For fossil energy resources, we see no possibility of ever becoming sustainable, the same applies to use of thorium and uranium in conventional nuclear power stations. However, if the technological and security challenges of breeder reactor designs and fuel recycling are overcome, then the perspectives for uranium and thorium may be widened to the order of 50,000 years.

From Figure 13 we can see a comparison of the different estimates with present recycling rates as repeated to the empirically based Hubbert-estimate. Making system dynamic models and running them may take some time to do. The Hubbert-estimate is fast, providing the right historical records are available. The plot of Hubbert-estimates versus the times obtained with dynamic modelling (Figure 13) shows good correlation; this means that we can make some observations and empirical rules:

1. Burn-off times multiplied by two is a good estimate for the time to scarcity. The burn-off estimate can be made on the back of an envelope, so that it is practical for quick calculations.
2. The Hubbert estimates of time to scarcity and system dynamic modelling estimates of time to scarcity give comparable results. The Hubbert method is a quicker way to get a rough estimate of time to scarcity.
3. System dynamic modelling estimates of time to scarcity makes a more detailed overview and allows for less sweeping generalizations and more specific adaptations to the situation of the specific element and the feedbacks that affects it's fate. Dynamic modelling allows for inclusion of systemic feedbacks and sensitivity analysis, as well as lending it self to policy optimization through back-casting.

It is therefore imperative that we start on a path towards sustainable development worldwide. Whether new technologies use components that will still be available in a few decades, should be a key criterion for their development, not an afterthought. Lack of resources is a dangerous situation globally; there are many convincing examples where this is indirectly or directly the cause for social crisis and potentially also war (Hardin, 1968, Bahn & Flenley, 1992, Ehrlich & Ehrlich, 1992, Haraldsson et al., 2002, 2007, Diamond, 2005, Klein, 2007, Lövin, 2007, Tilly, 2007, Zhang et al., 2007, Brown, 2009b). The solutions to our sustainability problems are as much in the social domain as in the technology domain. Engineering and economics must learn to deal with the functions and mechanisms of the social machinery, and realize that people and feedbacks from social processes control and shape human behaviour. Behaviour is what controls decisions, and these are not always conscientious, nor openly rational. The notion of "the rational economic man" as a foundation for economic behaviour is a mythomania, and it has never been shown to be

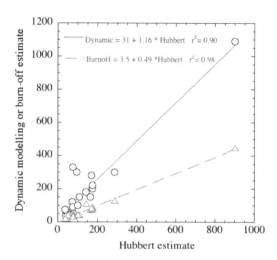

Fig. 13. The plot of Hubbert-estimates time to scarcity (years) versus the times obtained with system dynamic modeling (years) shows good correlation. The burn-off estimates are roughly 50% of the real time to scarcity, and thus underestimates the time to scarcity.

valid. The sustainability challenge is thus a social challenge and centres around the willingness of populations to change behaviour. The use of all resources available to us at maximum rate, probably possesses a threat or significant limitation to future generations, and carries large moral and ethical problems with them (Norgaard & Horworth, 1991, Costanza & Daly, 1992, MacIntosh & Edward-Jones, 2000, Ainsworth & Sumaila, 2003).

4. Conclusions

Our analysis of the length that resources will last into the future allows us to make the following conclusions:

1. Society is outrunning the global resource supply rate for key resources rapidly and has struck upon the planetary boundaries for resource supply for many materials.
2. With the prevailing one way-use paradigm, implying little or no recycling, the Earth cannot feed and sustain 7-9 billion people for very long. We show that there are some important end-times within 100-200 years from now, unless some paradigm changes have occurred.
3. The paradigm change includes policy-changes involving both convergence (efficiency, reduce losses, recycling) and contraction (population contraction, less intensive resource use, smaller extraction rates).
4. It will be possible to feed and supply approximately 2.5 - 3 billion people on Earth if we carefully recycle most of the resources (90-95% should be a target), making sure that we can keep enough material in the cycle, having low restocking demands, because of low losses.
5. Our bulk energy strategies are at present based on unsustainable thinking and still, inefficient use, and partly inadequate technologies.

a. We are currently in a chicken-race to consume all oil, gas and coal, with no honest thought for future generations. Many reserves are extracted very fast, involving poor recovery efficiencies.

b. For air transport there is at present no substitute fuel available, nor any alternative planned for. No trials are being undertaken. No viable technical proposal is on the drawing board. There is no plan B.

6. When it comes to public policies and strategic planning for the national states, a complete rethinking must take place in order to step out of a self-destructing behaviour. For approaching a sustainable situation in our world, recycling must be raised to levels between 80% and 95%; these are very challenging tasks both technologically and behaviourally.

It is important to realize that we cannot base any of our planning on miracles to happen in the future. When problems escalate towards crisis, we cannot assume that "someone smart" will show up with a magical solution that liberates us from all problems. Some researchers like to consider energy and material resources to be endless and unlimited, hoping for some yet undiscovered miracle. This is an inadequate attitude for future planning of sustainability, and there are many examples where such approaches failed in the past (e.g. Bahn & Flenley, 1992). At conferences we sometimes hear haggling over the exact numbers in the reserves tables, however this appears to the authors of this chapter as irresponsible, because doubling all our reserves estimates will not make the fundamental underlying problem go away. Neither does sustainability come around from talking about it; what we really need to do is to plan for real change to paradigms, social functions and human consumption patterns that allow us to become demonstrably sustainable. We must plan to become sustainable with the technologies that already exist and are available now, and with significant changes in societal and consumer behaviours. We must plan to change and set out with actions, even if it implies changing our behaviour, our lives, social standards for value, and our laws and policies. It now appears that unsustainability will soon challenge us, our daily lives and put democratic governmental form to tough tests. We need to get ready for real change, because if we do not then future generations do not have a chance to live on our Earth.

5. Acknowledgements

This chapter is an extended version of a commentary in Nature Geoscience in 2007 by Ragnarsdottir, in 2009 by Sverdrup et al., and two articles by Ragnarsdottir et al., 2011, and Sverdrup and Ragnarsdottir 2011. We are indebted to Colin Campbell, Bert de Vries, Martin Goldhaber, Charles Hall, Niels Mayer, Dennis Meadows, Jørgen Nørgård, Karl-Henrik Robért and Jane Plant for insightful discussions.

6. References

Adler, M. & Ziglio, E. (1996). Gazing Into the Oracle: The Delphi Method and its Application to Social Policy and Public Health. Kingsley Publishers: London.

Ainsworth, C. & Sumaila, U. (2003). Intergenerational discounting and the conservation of fisheries resources: A case study in Canadian Atlantic cod, In Sumaila U (ed.) Three

Essays on the Economics of Fishing. Fisheries Centre Research Reports Vancouver, pp. 26-33.

Arleklett, K. (2003). The peak and decline of world oil and gas production. Journal of Minerals and Energy, Vol. 18, pp. 5-20.

Arleklett, K. (2007). Peak oil and the evolving strategies of oil importing and exporting countries. Discussion paper 17. Joint Transport Research Centre – OECD - International Transport Forum.

Bahn, P. & Flenley, J. (1992). Easter Island, Earth Island. Thames and Hudson Ltd, London.

Bertalanffy, L. (1968). General System Theory: Foundations, Development, Applications. George Braziller, New York.

Brunner, P.H. and Rechberger, H. (2003). Practical Handbook of Material Flow Analysis. Lewis Publishers, London.

BP (2008). British Petroleum Statistical Review of World Energy, London.

Brown, L.R. (2009a). Plan B 4.0. Mobilizing to Save Civilization. Earth Policy Institute, Norton and Company, London.

Brown, L.B. (2009b). Could food shortages bring down civilization? Scientific American, Vol. 301, pp. 38-45.

Brown, L.R. & Kane, H. (1994). Full House: Reassessing the Earth's Population Carrying Capacity. Norton, New York.

Cavallo, A.J. (2004). Hubbert's Petroleum Production Model: An Evaluation and Implications for World Oil Production Forecasts. Natural Resources Research, Vol. 13, No. 4, December, 211-221.

Cavana, R.Y. & Mares, E.D. (2004). Integrating critical thinking and systems thinking: from premises to causal loops. System Dynamics Review, Vol. 20, pp. 223-235.

Cohen, J.E. (1995). Population Growth and Earth's Human Carrying Capacity. Science, Vol. 269, pp. 341-346.

Cohen, C., (2007). Earth's natural wealth, an audit. New Scientist May 23rd, Vol. 2605, pp. 34.

Costanza, R. & Daly, H.E. (1992). Natural capital and sustainable development. Conservation Biology, Vol. 6, pp. 1-10.

Cross, J. (2000). Gold derivatives: The market view Report from the World Gold Council London, p. 195.

Daily, G.C., Ehrlich, A.H. & Ehrlich, P.R. (1994). Optimum Human Population Size. Population and Environment, Vol. 15, pp. 469-475.

Daily, G.C., & Ehrlich, P. (1992). Population, sustainability and the earth's carrying capacity. A framework for estimating population sizes and lifestyles that could be sustained without undermining future generations. Journal of Bioscience, Vol. 42, pp. 761-771.

Diamond, J. (2005). Collapse, How Societies Choose to Fail or Survive. Penguin Books, London.

Eddington, Sir A.S. (1928). The Nature of the Physical World. MacMillan, 1935 replica edition. University of Michigan 1981 edition (1926–27 Gifford lectures).

Evans, L.T. (1998). Feeding the Ten Billion - Plants and Population Growth. Cambridge University Press, Cambridge.

Ehrlich, P.R. (1968). The Population Bomb. Ballantine books, New York.

Ehrlich, P. & Ehrlich, A. (1990a). The Population Explosion. Simon and Shuster, New York.

Ehrlich, P. & Ehrlich, A. (1990b). The population explosion. Why isn't everyone as scared as we are? The Amicus Journal, Vol. 12, pp. 22-29.

Ehrlich, P.R., Ehrlich, A.H. & Daily, G.C. (1992). Population, ecosystem services, and the human food supply. Morrison Institute for Population and Resource Studies Working Paper No. 44. Stanford University, Stanford, CA.

Ehrlich, P., Daily, G., & Goulder, L. (1992). Population growth, economic growth and market economics. Contentio, Vol. 2, pp. 17-35.

Ehrlich, P. & Ehrlich, A. (2006). Enough already. New Scientist, Vol. 191, pp. 46-50.

Ehrlich, P. & Ehrlich, A. (2009). The population bomb revisited. The Electronic Journal of Sustainable Development, Vol. 1, pp. 63-71.

Energy, Information Administration (2007). U.S. Crude Oil, Natural Gas, and Natural Gas Liquids Reserves, 2006 Annual Report. DOE/EIA-0216.

Evans, L.T. (1998). Feeding the Ten Billion - Plants and Population Growth. Cambridge University Press, Cambridge.

Filippelli, G.M., (2002). The global phosphorus cycle. In: Reviews in Mineralogy and Geochemistry, Vol. 48, Kohn, M., Rakovan, J., & Hughes, J. (eds.), pp. 91-125.

Fillippelli, G.M. (2008). The global phosphorus cycle: Past, present and future. Elements Vol. 4, pp. 89-95.

Forrester, J.W. 1961. Industrial Dynamics. Cambridge, MA: The MIT Press. Pegasus Communications, Waltham, MA.

Forrester, J. (1971). World Dynamics. Pegasus Communications, Waltham MA.

Francois, J.-L., Nunez-Carrera, A., Espinosa-Paredes, G. & Martin-del-Campo, C. (2004). Design of a boiling water reactor equilibrium core using thorium-uranium fuel. Proceedings from the American Nuclear Energy Symposium of 2004.

Friedman, M. & Friedman, R. (1980). Free to Choose. Harcourt, New York.

Friedman, M. (1962). Capitalism and Freedom. University of Chicago Press, Chicago.

Gordon, R.B., Graedel, T.E., Bertram, M., Fuse, K., Lifset, R., Rechberger, H. & Spatari, S. (2003). The characterisation of technological zinc cycles. Resource Conservation Recycling, Vol. 39, pp. 107–135.

Gordon, R., Bertram, B. & Grädel, M. (2006). Metal Stocks and Sustainability. Proceedings of the National Academy of Science, Vol. 103, pp. 1209-1214.

Greene, D.L., Hopson, J.L., Li, J. (2003). Running Out of and Into Oil: Analyzing Global Depletion and Transition Through 2050. Oak Ridge National Laboratory, Oak Ridge, Tennessee.

Goklany, I. (2009). Has increases in population, affluence and technology worsened human and environmental well-being? The Electronic Journal of Sustainable Development, Vol. 1, pp. 3-28.

Gott III, J.R. (1994). Future Prospects Discussed. Nature, Vol. 368, pp. 108-109.

Graedel, T. & Allenby, B. (1995). Industrial Ecology. Prentice Hall Publishers, Englewood Cliffs, New Jersey.

Graedel, T.E., Bertram, B., Fuse, K., Gordon R.B., Lifset, R., Rechberger, H. & Spatari, S. (2002). The contemporary European copper cycle; Characterization of technological copper cycles. Ecological Economics, Vol. 42, pp. 9–24.

Graedel, T., van Beers, D., Bertram, M., Fuse, K., Gordon, R. & Gritsinin, A. (2004). Multilevel Cycle of Anthropogenic Copper. Environmental Science and Technology, Vol. 38, pp. 1242-1252.

Greene, D.L., Hopson, J.L. & Li, J. (2003) Running Out of and Into Oil: Analyzing Global Depletion and Transition Through 2050. Oak Ridge National Laboratory, Oak Ridge, Tennessee, ORNL/TM-2003/259.

Guinée, J. ed. (2002). Handbook on Life Cycle Assessment: Operational Guide to the ISO Standards. Kluwer Academic Publishers, Heidelberg.

Hall, C., Lindenberger, D., Kummel, R., Kroeger, T. & Eichhorn, W. (2001). The need to reintegrate the natural sciences with economics. BioScience, Vol. 51, pp. 663–667.

Hall, C. (2008) Energy return on investment and our economic future. In Proceedings of the 33rd International Geological Congress, Oslo, Norway, August 2008.

Haraldsson, H., Sverdrup, H., Belyazid, S., Holmqvist, J. & Gramstad, R. (2007) The tyranny of small steps: A reoccurring behaviour in management. Journal of Systems Research and Behavioral Science, Vol. 24, pp. 1-19.

Haraldsson, H.V., Sverdrup, H., Svensson, M., Belyazid, S., Kalén, C. & Koca, D. (2002). The coming water shortage in the Jordan River Basin - Finding objectivity in a subjective problem. The 20th International Conference of the System Dynamic Society, Palermo, Italy, July 2002.

Haraldsson, H.V. (2004, 2005, 2006, 2007, 2008, 2009, 2010). Introduction to Systems Thinking and Causal Loop Diagrams. Reports in Ecology and Environmental Engineering. Department of Chemical Engineering, Lund University, Vol. 49, 7th Revised Edition.

Haraldsson, H.V. & Sverdrup, H.U. (2004). Finding simplicity in complexity in biogeochemical modelling. In: Environmental Modelling: A Practical Approach, Wainwright, J. and Mulligan, M. (Eds.) pp. 211-223. J. Wiley and Sons Ltd., Chichester.

Hardin, G. (1968) The tragedy of the commons. Science, Vol. 162, pp. 1243-1248.

Heinberg, R. (2007) Peak Everything: Waking Up to the Century of Decline in Earth's Resources, Clairview, London.

Hirsch, R.L., Bezdek, R. & Wending, R. (2005) Peaking of world oil production: Impacts, mitigation and risk management. "The Hirsch report" Available from at www.netl.doc.com/publ/other/pdf/oil_publ_net/pdf/

Hougen, O.A., Watson, K.M. & Ragatz, R. (1959) Chemical Process Principles. Part two: Thermodynamics. John Wiley and Sons, Inc., New York. .

Humphreys, M., Sachs, J. & Stiglitz, J. (2003). Escaping the Resource Curse. Columbia University Press, New York.

Hubbert, M. K. (1956). Nuclear energy and the fossil fuels. Presented to the spring meeting of the southern district, Division of Production. American Petroleum Institute. San Antonio, Texas. Publication no. 95, Shell Development company. Exploration and Productions Research Division, Houston, Texas.

Hubbert, M,K. (1966). History of Petroleum Geology and Its Bearing Upon Present and Future Exploration. AAPG Bulletin, Vol. 50, pp. 2504-2518.

Hubbert, M.K., (1972). Estimation of Oil and Gas Resources. In: U.S. Geological Survey, Workshop on Techniques of Mineral Resource Appraisal, pp. 16-50. U.S. Geological Survey, Denver.

Hubbert, M.K., (1982). Techniques of Prediction as Applied to Production of Oil and Gas, United States Department of Commerce, NBS Special Publication 631, May 1982.

International Copper Study Group (2004). Copper Bulletin Yearbook, Report, Lisbon, Portugal.

International Lead and Zinc Study Group (2003). Lead and Zinc Statistics. Monthly Bullulletin of the International Lead Zinc Study Group, Report 43, Lisbon.

International Nickel Study Group (2008). World Nickel Statistics. Monthly Bulletin Vol XVII, INSG Lisbon, Portugal.

Jackson, T. (2009) Prosperity Without Growth. Economics for a Finite Planet. Earthscan, London.

Jayaram, K.M.V. (1985). An overview of world thorium resoures, incentives for further exploration and forecast for thorium requiremetns in the near future. Prodceedings of a tehcnical committee meeting on utilization of thorium-based nuclear fuel: current status and perspectives. International Atomic Agency, Vienna, Austria, pp. 8-21.

Johnson Matthey Corporation (2008). Platinum Review Journals from the years 1996 to 2008.

Juang, M.H., Tsai, I. & Cheng, H.C. (2008). Formation of thin-film transistors with a polycrystalline hetero-structure channel layer. Semiconductor Science and Technology, Vol. 23, pp. 85-89.

Kasten, P.R. (1998) Review of the Radkowsky thorium reactor concept. Science and Global Security, Vol. 7, 237-269.

Klein, N. (2007). The Shock Doctrine: The Rise of Disaster Capitalism. Knopf Canada Publishers, Mississauga, Ontario, Canada.

Korb, K. & Oliver, J. (1998). A refutation of the doomsday argument. Mind, Vol. 107, pp 403–410.

Leslie, J. (1998). The End of the World: The Science and Ethics of Human Extinction. Routledge, New York, NY, USA.

Lövin, I. (2007). Tyst Hav. Ordfront Förlag AB, Stockholm.

MacIntosh, A. & Edwards-Jones, G. (2000). Discounting the children's future? Geophilos, Vol. 1, pp. 122-133.

Malthus, T.R. (1798). An essay on the principle of population as it affects the future improvement of society, reprinted as "The mathematics of food and population". In: The World of Mathematic, Newman J. (ed.), Simon and Schuster, New York 1956, pp. 1192-1199.

Maani, K. & Cavana, R. (2000). Systems Thinking and Modelling – Understanding Change and Complexity, Prentice Hall, Auckland.

Mao, J. S., Dong, J. & Graedel, T.E. (2008). The multilevel cycle of anthropogenic lead II. Results and discussion, Resource Conservation and Recycling, Vol. 52, pp. 1050-1057

McGarvey, B. & Hannon, B. (2004). Dynamic Modelling for Business Management. Springer, New York, USA.

Meadows, D. H., Meadows, D. L. Randers, J. & Behrens, W. (1972). Limits to Growth. Universe Books, New York, USA.

Meadows, D.H, Meadows, D.L. & Randers, J. (1992). Beyond the Limits: Confronting Global Collapse. Envisioning a Sustainable Future. Chelsea Green Publishing Company, Chelsea.

Meadows, D.H., Randers, J., Meadows, D. (2004). Limits to Growth. The 30 Year Update. Universe Press, New York, USA.

Meridian International Research (2008). The trouble with lithium 2. Under the microscope. Meridian International Research Report.

Norgaard, R.B. & Horworth, B., (1991). Sustainability and discounting the future. In: Ecological Economics: The Science and Management of Sustainability, Costanza, R. (Ed.) 1991. Columbia University Press, New York, USA.

OCC (2009). Controller of the Currency Administrator of National Banks. Washington, DC 20219 OCC's Quarterly Report on Bank Trading and Derivatives Activities, Fourth Quarter 2009.

Oelkers, E.H. & Valsami-Jones, E. (2008). Phosphate mineral reactivity and global sustainability. Elements, Vol. 4, pp. 83-87.

Osborn, F.L. (1948). Our Plundered Planet. Pyramid Publications, New York, USA.

Pearson, F.A. & Harper, F.A. (1945). The World's Hunger. Harper Cornell Univ. Press, Ithaca, NY, USA.

Penck, A. (1925). Das Hauptproblem der Physischen Anthropogeographie. Zeitschrift für Geopolitik, Vol. 2, 330-348.

Pogue, J.E. & Hill K.E. (1956). Future Growth and Financial Requirements of the World Petroleum Industry. Presented at the annual meeting of the American Institute of Mining, Metallurgical, and Petroleum Engineers, Petroleum Branch, New York, New York, February 1956.

Ragnarsdottir, K.V. (2008). Rare metals getting rarer. Nature Geoscience, Vol. 1, pp. 720-721.

Ragnarsdottir, K.V., Sverdrup, H. & Koca, D. (2011). Challenging the planetary boundaries I: Basic principles of an integrated model for phosphorus supply dynamics and global population size. Applied Geochemistry, Vol. 26, pp. S307-310.

Rockström, J., Steffen, W., Noone, K., Persson, A., Chapin III, F.S., Lambin, E., Lenton, T.M., Scheffer, M., Folke, C., Schellnhuber, H., Nykvist, B., De Wit, C.A., Hughes, T., van der Leeuw, S., Rodhe, H., Sörlin, S., Snyder, P.K., Costanza, R., Svedin, U., Falkenmark, M., Karlberg, L., Corell, R.W., Fabry, V.J., Hansen, J., Walker, B.H., Liverman, D., Richardson, K., Crutzen, C., Foley, J. (2009). A safe operating space for humanity. Nature, Vol. 461, pp. 472–475.

Roskill Information Services Ltd. (2007a). The Economics of Rare Earths and Yttrium, Sydney, Australia.

Roskill Information Services Ltd. (2007b). The Economics of Tungsten, 9th Edition, Sydney, Australia.

Roskill Information Services Ltd. (2007c). The Economics of Cobalt, 11th Edition, Sydney, Australia.

Roskill Information Services Ltd. (2007d). The Economics of , 10th Edition, Sydney, Australia.

Roskill Information Services Ltd. (2008). The Economics of Bauxite, 7th Edition, Sydney, Australia.

Roskill Information Services Ltd. (2009a). The Economics of Tantalum, 10th Edition, Sydney, Australia.

Roskill Information Services Ltd. (2009b). The Economics of Niobium, 11th Edition, Sydney, Australia.

Roskill Information Services Ltd. (2010a). Titanium Metal: Market Outlook to 2015, 5th Edition, Sydney, Australia.

Roskill Information Services Ltd. (2010b). Indium: Global industry markets and outlook, 9th Edition, Sydney, Australia.

Roskill Information Services Ltd. (2010c). Rhenium: Market outlook to 2015, 8th Edition, Sydney, Australia.

Roskill Information Services Ltd. (2011). Gallium: Global Industry Markets and Outlook, 8th Edition, Sydney, Australia.

Råde, I. (2001). Requirement and Availability of Scarce Metals for Fuel-cell and Battery Electric Vehicles. PhD thesis, Chalmers University, Gothenburg, Sweden.

Sachs, J. (2008). Common Wealth: Economics for a Crowded Planet. Penguin Press. London, UK.

Senge P. (1990). The Fifth Discipline, The Art and Practice of the Learning Organisation. Century Business, New York, NY, USA.

Shafik, N. (1994). Economic development and environmental quality, an econometric analysis. Oxford Economic Papers, Vol. 46, 757-773.

Smil, V. (2001). Enriching the Earth: Fritz Haber, Carl Bosch, and the Transformation of World Food Production. MIT Press, Cambridge, MA, USA.

Smil, V. (2002). Phosphorus, global transfers. Causes and consequences of environmental change. In: Encyclopedia of Global Environmental Change, Douglas, I. (ed), pp. 536-542. John Wiley and sons, Chichester, UK.

Smith, A. (1776). An Inquiry into the Nature and Causes of the Wealth of Nations, Reprint by the University of Chicago Press, Chicago, IL, USA.

Sober, E. (2003). An Empirical Critique of Two Versions of the Doomsday Argument - Gott's Line and Leslie's Wedge, Synthese, Vol. 135-133, pp. 415-430.

Sowers, G.F. (2002). The demise of the doomsday argument. Mind, Vol. 111, pp. 37-45.

Spatari, S., Bertram, M., Gordon, R. B., Henderson, K. & Graedel, T. E. (2005). Twentieth century copper stocks and flows in. North America: A dynamic analysis. Ecological Economics, Vol. 54, pp. 37–51.

Strahan, D. (2007). The Last Oil Shock: A Survival Guide to the Imminent Extinction of Petroleum Man. John Murray Books, New York, NY, USA.

Strahan, D. (2008). Coal: Bleak outlook for the black stuff. New Scientist, Vol. 2639, pp. 38-41.

Sterman, J.D. (2000). Business Dynamics, System Thinking and Modeling for a Complex World. Irwin McGraw-Hill, New York, NY, USA.

Sverdrup, H. & Svensson, M. (2002a). Defining Sustainability in Developing Principles for Sustainable Forestry; Results from a Research Program in Southern Sweden. In: Managing Forest Ecosystems, Sverdrup, H. & Stjernquist, I. (ed.), pp. 21-32. Kluwer Academic Publishers, Amsterdam, Netherlands.

Sverdrup, H. & Svensson, M. (2002b). Defining the Concept of Sustainability, a Matter of Systems Analysis. In Revealing Complex Structures - Challenges for Swedish Systems Analysis. Sjöstedt M. (ed.), pp. 122-142. Kluwer Academic Publishers, Amsterdam, Netherlands.

Sverdrup, H., Koca, D. & Robèrt, K.H., (2009). Towards a world of limits: The issue of human resource follies. Geochimica et Cosmochimica Acta, Vol. 73, pp. A1298.

Sverdrup, H. and Ragnarsdottir, K.V. (2011). Challenging the planetary boundaries II: Assessing the sustainable global population and phosphate supply, using a systems dynamics assessment model. Applied Geochemistry Vol. 26, pp. S311-S314.

Tans, S. J., Verschueren, A. R. M. & Dekker, C. (1998) Room temperature transistor based on a single carbon nanotube. Nature, Vol. 393, pp. 49-52.

Tilly, C. (2007). Democracy. Cambridge University Press, Cambridge, UK.

Turner, G. (2008). A comparison of the limits to growth with thirty years of reality. Socio-economics and the environment in discussion: CSIRO Working papers, Series 2008-2009.

USGS (2002) 1998 Assessment of Undiscovered Deposits of Gold, Silver, Copper, Lead, and Zinc in the United States, Circular 1178 (U.S. Geol. Survey, Washington, DC).

USGS (2008). Commodity Statistics for a Number of Metals (Consulted Several Times 2008-2010). United States Geological Survey.
http://minerals.usgs.gov/ minerals/pubs/commodity/.

Vennix, J.A.M, Andersen, D., Richardson, P.G. & Rohrbaugh, J. (1992). Model-building for group decision support: Issues and alternatives in knowledge elicitation. European Journal of Operational Research, Vol. 59, pp. 14-28.

von Liebig, J., Playfair, L., Webster, J.W. (1841). Organic Chemistry and its Applications to Agriculture and Physiology. J. Owen Publishing House, Cambridge.

von Liebig, J. (1843). Familiar Letters of Chemistry and its Relation to Commerce, Physiology and Agriculture. J. Gardener, London.

Williams, N. (2008). The exhaustion of mineral resources: a truism or a state of mind? Proceedings of the 33rd International Geological Congress, Oslo, August 2008.

Zhang, D., Zhang, J., Lee, H.F. & He, Y. (2007). Climate change and war frequency in eastern China over the last millennium. Human Ecology, Vol. 35, pp. 403-414.

Zittel ,W., & Schindler, J. (2007). Coal: Resources and Future Production. A background paper produced by the Energy Watch Group. Ludwig Bolkow Foundation, Ottobrunn, Germany.

Part 4

Sustainable Environment and Water Management

7

Integrated Water Resources Management - Key to Sustainable Development and Management of Water Resources: Case of Malawi

V. Chipofya[1], S. Kainja[2] and S. Bota[2]
[1]University of Malawi, The Polytechnic,
[2]Malawi Water Partnership Secretariat,
c/o Malawi Polytechnic,
Malawi

1. Introduction

1.1 Country's location and surface area

Malawi lies between latitudes 9°S and 17°S and between longitudes 33°E and 36°E. The country's international frontiers are shared with the United Republic of Tanzania to the north and northeast, the Republic of Mozambique to the southeast, south and southwest and the Republic of Zambia to the west. It covers a geographical area of 118,484 km². Lake Malawi, the third largest freshwater lake on the African continent, takes up nearly 23.6 percent or 28,000 km² of the area. Malawi is a founding member of Southern Africa Development Community (SADC).

1.2 Population

The population of the country is estimated at 12 million, with a population density of 107 people/km². Most of the people (90 percent) live in rural areas. The main economic base of the country is agriculture with subsistence and smallholder farming being prevalent among the rural population.

1.3 Natural resources

The country has a range of natural resources which include fertile soils for agriculture, water resources, and a remarkable diversity of flora and fauna that have earned Malawi a unique habitat for bio-diversity. However, these natural resources are continuously threatened by high population densities and poverty, which have led to widespread deforestation, cultivation and settlement of marginal areas for survival. These factors highlight the challenge of balancing efforts between poverty alleviation (economic growth) and natural resources management.

1.4 Climate

The climate of Malawi is influenced by the country's geographical location. Lying northward of the sub-tropical high-pressure belt, the country is affected by south-easterly winds for about six months of the year. The dominant wind system influencing the country's climate is the position of the Inter-Tropical Convergence Zone (ITCZ), which oscillates north and south bringing with it the changes in seasons as it moves (see Figure 1). Thus, when there is strengthening of the south easterlies towards the ITCZ, which normally lies over the Central Region of the country, increases in cloud cover occur resulting in rainfall. Local topography also determines climatic conditions. Due to the diverse topography that Malawi has and the great range in altitude between locations, climatic conditions may be complex. Variations between wet and dry places and between hot and cold areas are therefore not uncommon due to this characteristic.

Temperature

The period between May and August is generally characterised by low temperatures and relative humidity. With the advance of the rainy season in about October until January, temperatures are usually high and humidity also increases. A greater part of the country enjoys favourable and tolerable temperatures especially on the plateau areas, which register moderately low temperatures as compared with the Lakeshore and the Lower Shire Valley. The mean maximum July temperature for a larger part of the west Mzimba plains, the Central Region plateau and the area extending from the southern boundary of the Thyolo escarpment to Lake Chilwa and beyond to Namwera is between 22.5° and 25° with an approximate range of between 10° to 12.5°. During this time, the high plateau areas of the Nyika, Viphya, Dedza, Zomba and Mulanje may record mean maximum temperatures of between 12.5° and 15°C with a range of about 5° to 10°C. Along the Lakeshore and in the Lower Shire Valley, mean maximum July temperatures are usually higher than 35°C.

Rainfall

A number of rainfall measuring stations exist in the country and are run by the Department of Meteorological Services as well as others including the Department of Forestry, Ministry of Agriculture, Department of Parks and Wildlife, Ministry of Irrigation and Water Development and the private sector. The direction of the prevailing winds has an important influence on the amount of rainfall received. For instance, the tangential incidence of the south-easterly winds on the western shores of Lake Malawi brings with it high rainfall around these areas. However, if the direction of the winds is parallel to the orientation of the shore, this results in no or little rainfall. The highest rainfall in the country occurs around the area north of Karonga Boma with intensities of higher than 2,050mm as well as around Nkhata Bay, Nkhotakota, Zomba and the south-eastern corner of the country in Thyolo and Mulanje. A steep southerly gradient of rainfall intensity is evident from Mwangulukulu, which rises again upon approaching the Nyika Plateau.

March is the wettest month in the year. Similarly, the great diversity of topography in Malawi sees those areas on the windward side receiving much higher rainfall than those on the leeward side, with Nkhata Bay receiving mean annual rainfall of above 1,850mm and Mzimba having only around 820mm to 1,030mm. The lowest rainfall is received in areas of low altitude such as in the Lower Shire Valley where the mean annual rainfall is below 820mm.

Integrated Water Resources Management - Key to Sustainable Development and Management
of Water Resources: Case of Malawi

147

2. Enabling environment

2.1 Legislative provisions for water management

The need for sustainable development and management of water resources in Malawi is underscored by the existing policy guidelines, institutional arrangements and regulatory framework. These regulatory instruments are aimed at safeguarding the ecologically fragile and sensitive receiving water courses where the water, further downstream is used by people for washing clothes and bathing, or irrigating crops which may be eaten raw (Carl Bro International, 1995).

A number of water management policies and legislations have been enacted in the country. The policies and legislations have been regulatory in nature. The Water Resources Act (1969) and its subsidiary Water Resources (Pollution Control) Regulations provide the main regulatory framework for water resources management. On the other hand, Water Works Act (1995) is the main authority that established water supply and water borne sanitation delivery services. There is a high degree of policy harmonization and collaboration amongst institutions dealing with water and environmental sanitation in Malawi (Chipofya et al., 2009).

The National Water Policy (NWP) (2005) ensures water of acceptable quality for all needs in Malawi.

The National Sanitation Policy (NSP) (2008) stipulates the need to improve delivery of improved sanitation services.

Further to the above policy framework relating to water pollution control, the Malawi Government launched the Malawi Growth and Development Strategy (MGDS) in 2007. The MGDS is the overarching operational medium term strategy for Malawi designed to attain the nation's Vision 2020 (1995).

One of the nine priority areas in the MGDS is Irrigation and Water Development. Under this priority area is a sub-theme for conservation of the natural resource base and in particular water supply and sanitation.

In addition, formalized national effluent standards exist in Malawi (MBS, 2005). The main policing agent to ensure compliance is the Department of Environmental Affairs in the Ministry of Natural Resources, Energy and Environment.

Malawi, as a member state of the United Nations (UN), is also obliged to meet the UN Millennium Development Goals (MDGs) www.un.org/millenniumgoals/ (accessed 09.02.2010). Goal number seven in the MDGs relates to ensuring environmental sustainability by 2015.

2.2 National development strategies and implications for water resources management

Malawi's strategy on Water Resources Management is aimed at improving water resources conservation and storage through flow regulation, promotion of small and medium to large dams, and exploitation of ground water to meet the country's target of poverty reduction.

The Malawi Vision 2020, a national long-term Development Perspective, articulates the aspirations of Malawians and the development prospects of the country up to the year 2020.

This document, among other things, showcases the threats to the country's water resources and the environment, and also offers solutions for long-term protection and utilization. The vision recognizes that water is a limited and essential resource that is sometimes taken for granted particularly among the urban dwellers. In the rural areas of Malawi, however, people are confronted directly by its elusive nature as they are vulnerable to the ravaging cycles of drought and floods, and the slowly degrading resource base.

Malawi's long-term development goals embedded in the Vision 2020 have been translated into implementable medium term strategies through Malawi Development and Growth Strategy (MDGS). The MDGS recognizes the importance of water and assumes that there will be adequate water resources as seen in the following economic growth priorities:

Agriculture is the backbone of the economy contributing 63.7 percent of the total income of the rural people, 36 percent of GDP, 87 percent of total employment and supplying more than 65 percent of the manufacturing sector's raw materials. Agriculture has been prioritized as one of the high growth sectors in the economy's growth strategy. Increased agriculture production will be achieved through increased use of water using irrigation farming for both smallholder and commercial farming and increased use of fertilizers. If not well managed, the down side of this growth strategy would be drastic reduction in quantity and quality of water for use by other development uses due to increased water demand and increased water pollution by fertilizers and agrochemicals. This would threaten long-term sustainable economic growth.

Agro-processing is another priority area in the medium term. Also earmarked for growth is industrial processing. Adoption of more intensive production and processing methods will lead to production of large quantities of solid and liquid waste and discharges that have the potential to pollute both surface and ground water resources.

The economic growth strategy will require increasing usage of electricity which is currently mostly coming from hydroelectric sources. There are plans to rehabilitate the old hydropower generation stations and develop new ones in potential rivers such as Songwe, North Rukuru, South Rukuru and Ruo rivers. Sustenance of power generation will require maintaining steady flows in the rivers which entails proper management of the water resources upstream of the stations.

Malawi is working towards reducing by half, in the short term, the number of people without access to clean water and good sanitation, and providing good water and sanitation to every Malawian by 2025. Currently, only about 65 percent of the people have access to safe water. As the population grows, attainment of this objective entails nearly doubling the current efforts of supplying clean water in all areas of the country (urban, peri-urban, and rural).

The planned development of the Zambezi Waterway is envisaged to bring economic growth to the country through reduced transport costs and increased tourist activities. For this massive economic undertaking to be sustained and bear the required fruits in the medium and long term, water resources management to maintain a steady flow rate in the Shire River will be a requirement. Water pollution control will be critical as well as trans-boundary issues.

Mining and Tourism sectors are also earmarked as priority economic growth areas. Mining operations require substantial amounts of water which results into effluent and solid waste that can degrade the quality of water available down stream. Tourism on the other hand, is heavily concentrated along the lakeshore areas, necessitating provision of good sanitation. Health risks along the lake such as threat of contracting bilharzia or malaria both of which are water related diseases could be detrimental to the industry if not addressed adequately.

The development of aquaculture involves increased use of fishponds for fish production and depends on clean and uncontaminated water quality. Plans to intensify fish production in lakes, dams, rivers and fishponds are dependent upon availability of adequate clean water.

The Malawi Growth and Development agenda is essentially an agenda to use more and more water. Sustainable development for Malawi calls for adoption of an integrated water resources management strategy if the water crisis projected to occur in 2025 is to be avoided.

Implementation of strategies identified in the MGDS will also enable Malawi to achieve internationally set targets such as the Millennium Development Goals (MDGs). All these documents contain goals and targets that among other things aim at reducing substantially the number of people living in poverty, improving access to the basic human needs (enough food, basic education and basic health care) and sustainable management of the environment.

The core problem facing the water sector is the challenge of maintaining a balance between exploitation of water resources for social economic development and sustainable management of the resources. Achievement of the twin objectives is possible through integrated water resource management.

2.3 Key challenges for the water sector

Water resource management challenges can be grouped into two categories namely: those associated with natural systems and those associated with human systems. Natural systems challenges constitute floods and drought mitigation. Both challenges are caused by climate change and climate variability which in turn is exacerbated by global warming. Malawi faces frequent floods of which the more recent ones occurred in the 1991/1992 and 1994/1995 rain seasons. In addition, there are increasing frequencies of floods especially along the lakeshore and Shire River system. Floods cause a lot of damage to property and loss of lives of many people every year. The current response to flooding is however, reactive. There are no mitigation or adaptation measures yet in place.

Human systems that have been established to address a number of water and water-related challenges are not functioning effectively, resulting in concerns such as poor catchment management, low capacity for IWRM/WE implementation, poor stakeholder coordination, poor information management systems, low maintenance of water delivery systems, and water quality degradation. Some human systems failures are associated with cross-cutting concerns, such as HIV/AIDS and Gender, which impact on water resource management.

2.4 Water sector reform process

Prior to the 1980s, the water sector received relatively little investment in infrastructure development or water resources management. In the 1980s, mainly as a result of the international water supply and sanitation decade initiative, more attention was given to the water sector which resulted in a substantial increase in investment in the sector. More water points were provided through boreholes, shallow wells and gravity-fed piped water supply schemes in rural areas. In cities and towns, improved access was made through investments by Water Boards and District Water Supply Fund (DWSF).

The water sector services study of 1993/94 analyzed the water sector and identified weaknesses such as lack of coherent policy framework, weak legal instruments, inappropriate institutional arrangements, lack of capacity, and inappropriate strategies for service provision. These findings culminated in the development of the first phase of the National Water Development Project, hereafter referred to as NWDP (I), which was seen as a vehicle towards implementing the water policy and other recommendations in the study. The main objective of the project was to support the implementation of the 1994 policy to ensure adequate and safe water supply, provision of water infrastructure, and protection and management of water resources. The main outputs of NWDP (I) were:

- Creation of the 3 Regional Water Boards;
- Completion of 6 water resource management and development studies relating to Lake Malawi and other major rivers and catchment areas;
- Construction of a dam and water supply systems in the Municipality of Zomba and 17 other towns;
- Development of a district-based community managed approach to rural water and sanitation;
- Construction of 500 boreholes and 2 gravity piped water supply schemes; and
- Capacity building in the Ministry of Irrigation and Water Development and the five Water Boards (NRWB, CRWB, SRWB, BWB and LWB) through the provision of training, equipment and technical assistance.

The implementation of the NWDP (I) brought about some improvements in the water supply and sanitation delivery and water resources management. However, some short falls still remain. NWDP (II) has therefore been developed to address the shortcomings of NWDP (I). This phase will build on the experiences and achievements of the first phase, consolidate the sector institutions, improve on water resources management and accelerate the provision of water and sanitation services to the communities in a sustainable manner. The objective of NWDP (II) is to improve water resources management and increase access to sustainable water supply and sanitation services for the people living in cities, towns and villages. Its main components are:

- Urban water supply and sanitation in Blantyre and Lilongwe Water Boards;
- Town water supply and sanitation in the three regional water boards;
- Rural water supply and sanitation in District Assemblies; and
- Water resources management.

3. Water resource situation in Malawi

3.1 Spatial and seasonal distribution of surface water

Malawi has a good network of river systems and is rich in surface water resources. Some of the water systems are shared with neighbouring countries of Tanzania and Mozambique, and on wider scale form part of the Zambezi River Basin. Most (93.2 percent) of Malawi's

Integrated Water Resources Management - Key to Sustainable Development and Management
of Water Resources: Case of Malawi

151

territorial area is in the Zambezi Basin and 86.1 percent of her population live in the Basin. However, a lot of imbalances exist in the spatial and seasonal distribution of surface water. Relatively few areas have abundant water resources available throughout the year, with most areas experiencing seasonal fluctuations or perpetual year to year water scarcity with pronounced shortages during the dry months of the year. Most rivers dry up by July, with the exception of those flowing from high altitude rainfall areas of Nyika and Viphya plateaux, the Kirk Mountain Ranges, Zomba Plateau and Mulanje Mountain. This situation of unreliable dry season flows has been exacerbated by deforestation and land use malpractices which, together with improper and in some cases unwarranted usage of heavy agrochemicals and unchecked disposal of domestic and industrial wastes and allied effluent matter, have substantially deteriorated the surface water resources. These have particularly occurred in headwaters, escarpments, and mountainous catchment areas which are, in normal circumstances, supposed to exist as protected land.

The drainage system of Malawi has been divided into 17 Water Resources Areas (WRA) and each WRA represents one basin. The WRAs are sub-divided into 78 Water Resources Units. Lake Malawi stores the bulk of the renewable surface water resources. The Shire is the largest river and it is the only outlet of Lake Malawi, while all the other major rivers drain into Lake Malawi or Shire River. A few of these major rivers drain into Lake Chilwa which is not part of the Lake Malawi catchment and therefore not part of the Zambezi River Basin. The Shire flows into the Zambezi in Mozambique.

Droughts and floods are recurrent in Malawi. The impact of climate change and variability strongly influences the occurrence and distribution of floods and droughts. The late start of the 2005/2006 rainfall season and inadequate rainfall during the season resulted in a dwindling of water resources. This was clearly evident in surface water resources as many rivers have had lower flows in the past water year. Even the lake levels have experienced a significant drop. For example, the mean lake level for October 2006 was 474.65 m.a.s.l. for October 2004. In the case of flooding, areas that are flood prone in Malawi include the Lower Shire Valley, lakeshore areas of Lake Malawi, Lake Chilwa and Lake Malombe. Of particular significance is flooding in the Lower Shire, which creates both national and regional problems since the river flows into Zambezi in neighbouring Mozambique.

3.2 Water resources availability and distribution

Malawi is a water stressed country with total renewable water resources per capita of less than 1400m³. With such low per capita, Malawi is worse of than Botswana and Namibia, countries which have large areas of desert and are traditionally believed to be water stressed.

a. Surface Water Resources

Malawi, in view of the large lake, high plateau and rugged relief, has a distinct climate. The country experiences good rainfall from November to April. The mean annual rainfall is 1,037mm. The rainfall distribution for the country is also varied according to altitude as shown in Table 1 below.

The mean monthly temperature ranges from 10° to 16°C in the highlands, 16° to 26°C in the plateau areas, 20° to 29°C along the lakeshore, 21° to 30°C in the Lower Shire Valley.

Range (mm)	Area (Km²)	Percentage Proportion
650 – 1,000	59,464km²	63.1
1,000 – 1,200	16,095	17.1
> 1,200	18,717	19.8

Table 1. Rainfall Distribution

The mean annual pan evaporation ranges from 1,500 – 2,000mm in the plateau areas and is highest (2,000 – 2,300mm) along lakeshore and Shire Valley. Lake Malawi stores the bulk of the renewable surface water resources with an average of 90Km³ of live storage (that can flow out of the Shire River). This lake, which is the third largest in Africa, has a surface area of 28,760 km² and an estimated total volume of water of $7,725 \times 10^9 m^3$ with a mean level of 474 m.a.s.l. It is the most important water resource for Malawi and plays a vital role in the socio-economic development of the country. The Shire River itself transits an annual average of about 18 Km³ (500 to 600 m³/s) into Mozambique. The annual surface water resources yields on land are about 13 Km³ and predominantly drain into Lake Malawi and the Shire River. However, more than 90 percent of this runoff occurs in rainy season, particularly from December to April every year.

Malawi, though with the largest riparian area of 65.9 percent, contributes only 42 percent of the total inflow into the lake, much less than Tanzania which, with only a riparian area of 27.2 percent, contributes about half of the inflow into the lake. This entails that Malawi needs the compliments of other riparian countries when managing and developing the lake resources. Table 2 gives the inland runoff contribution to the lake from three riparian countries of Malawi, Mozambique and Tanzania. This, therefore, calls for closer bilateral cooperation among these riparian countries.

	Malawi	Tanzania	Mozambique	Total
Catchment area of lake km²	64 372	26 600	6 768	97 740
Flow (m³/ s)	391	486	41	918
Percentage area (%)	65.9	27.2	6.9	100
Percentage flow (%)	42.6	52.9	4.5	100

Source: Ministry of Water Development

Table 2. Contribution of Run-off water to Lake Malawi

Major rivers are the Shire, Bua, Linthipe, Songwe, North Rukuru, South Rukuru, Dwangwa and Ruo. The Shire is the largest river and is the only outlet of Lake Malawi, while all the other major rivers drain into Lake Malawi or Shire River. The major catchment areas are as shown in Figure1.

The major rivers are perennial, but due to the seasonality of rainfall, most of the smaller rivers have ephemeral flow. The mean annual runoff works out to 588m³/s or 18,480 x 106m³. The mean annual runoff over the land area of the whole country is 196mm (i.e. an equivalent of 588 m³/s), and this constitutes 19 Percent of the mean annual rainfall. Details of rainfall and runoff for each WRA are shown in Table 3.

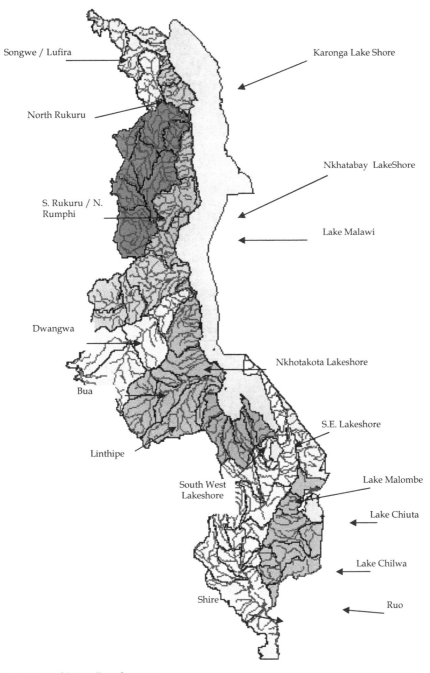

Source: Ministry of Water Development

Fig. 1. Malawi Water Resources Catchments

River Basin	Catchment Area Km²	Rainfall Mm	Runoff mm	Runoff M³/s	Percentage Runoff
Shire	18,945	902	137	82	15.2
Lake Chilwa	4,981	1,053	213	34	20.2
South West Lakeshore	4,958	851	169	27	19.9
Linthipe	8,641	964	151	41	15.7
Bua	10,654	1,032	103	35	10.0
Dwangwa	7,768	902	109	27	12.1
South Rukuru	11,993	873	115	44	13.2
North Rumphi	712	1,530	674	15	44.1
North Rukuru	2,091	970	252	17	26.0
Lufira	1,790	1,391	244	14	17.5
Songwe	1,890	1,601	327	20	20.4
South East Lake Shore	1,540	887	201	10	22.7
Lake Chiuta	2,462	1,135	247	19	21.8
Likoma Island	18.7	1,121	280	-	-
Chisumulo Island	3.3	1,121	280	-	-
Ruo	3,494	1,373	538	60	39.2
Nkhotakota Lakeshore	4,949	1,399	260	41	18.6
Nkhata Bay Lakeshore	5,458	1,438	461	80	32.1
Karonga Lakeshore	1,928	1,208	361	22	35.1
TOTAL	94,276	1,037	196	588	18.9

Source: Ministry of Water Development

Table 3. River Basins of Malawi: Mean Annual Rainfall and Runoff

Other important surface water resources include Lake Chilwa with a surface area of 683 km², Lake Malombe with 303 km², and Lake Chiuta with 60 km². Small lakes, lagoons and marshes include Lake Kazuni, Chia Lagoon, Chiwondo Lagoon, Elephant Marsh, Ndindi Marsh and Vwaza Marsh. Details are outlined in Table 4.

Reservoir	Surface Area (Km²)	Location as per District
Lake Malawi	28,750	Covers Karonga, Rumphi, Nkhatabay, Nkhotakota, Salima, Dedza & Mangochi
Lake Chilwa	683	Zomba & Phalombe
Lake Malombe	303	Mangochi
Lake Chiuta	60	Machinga
Lake Kazuni*	-	Rumphi & Mzimba
Chia Lagoon	22	Nkhaota kota
Chiwondo Lagoon	-	Karonga
Elephant Marsh*	-	Chikwawa & Nsanje
Ndindi Marsh*	-	Nsanje
Vwaza Marsh*	-	Rumphi

Source: Ministry of Water Development * Surface area not known

Table 4. Major Natural Reservoirs and Marshes in Malawi

b. Reservoirs

In Malawi, water resources utilisation is heavily dependent on run-of-the-river schemes. Although there is great potential and need for dams, no major storage dams have been constructed. However, there are about 700 small to medium dams that have been constructed with reservoir capacities ranging from a few cubic metres to about 5 million cubic metres. The total storage of these dams is estimated at about 100 million cubic metres or 0.1km³.

There are a total of 749 impoundments in the country, the majority (over 60 percent) of which are in South Rukuru and Ruo River basins. Most of these dams were constructed in the 1950's mainly to supply drinking water for livestock. The dams that can be classified as large dams in Malawi have mainly been developed by water boards for urban water supply. These include Lunyangwa Dam in Mzuzu, Chitete Dam in Kasungu, Kamuzu I and II Dams in Malingunde (Lilongwe), Mpira Dam in Ntcheu, Mulunguzi Dam in Zomba, Mudi and Chimwankhunda Dams in Blantyre, Chilingali Dam in Nkhotakota, and Lifupa Dam in Kasungu, as detailed in Table 5.

Water Resources Area	Number of Dams	Remarks
Shire	62	Most in Blantyre
Lake Chilwa	31	Most in Thondwe area
South West Lakeshore	8	
Linthipe	33	
Bua	38	
Dwangwa	50	
South Rukuru/North Rumphi	274	
North Rukuru	2	
Songwe / Lufira	3	
South East lakeshore	1	
Lake Chiuta	2	
Likoma Island	0	
Chizumulo Isalnd	0	
Ruo	215	
Nkhotakota Lakeshore	9	
NKhata-bay Lakeshore	21	Almost all in Luweya/Limphasa area
Karonga Lakeshore	0	
Total	749	

Source: Ministry of Water Development

Table 5. Distribution of Dams per River Basin

c) Groundwater Resources

Evaluation and development of groundwater resources have been primarily for drinking water supply for both rural and urban areas. The construction of boreholes and hand-dug wells, which started in the 1930's, can be considered to be the beginning of the utilisation of groundwater resources in Malawi. Basically there are two types of aquifer systems in the country namely the extensive but low yielding weathered basement aquifer of the plateau area, and the high yielding alluvial aquifer of the lake shore plains and the Lower Shire Valley. The weathered zone is best developed over the plateau areas where it is commonly 15 – 30 metres thick and even thicker. The average yield in the weathered zone of the basement complex lies in the range of 1 - 2 litres per second.

The alluvial aquifers are fluvial and lacustrine in nature, and are highly variable in character both in vertical sequence and lateral extent. They occur in several basins which, apart from Lake Chilwa areas, are all located along the rift valley floor: Karonga Lake Shore, Salima - Nkhotakota Lake Shore, Upper Shire Valley and the Lower Shire Valley. Most lithological records from boreholes give little detailed information about the successions. The overall impression is that clays usually dominate the sequence although in many localities there is significant thickness of poorly sorted sands. The sedimentary environments likely to produce the highest groundwater yields are buried river channels and littoral zones of the lake shore where the deposits are usually coarse grained and well sorted. The Lake Chilwa Basin is different from the other alluvial areas in that it is perched on the eastern side of the rift valley. The lithological logs of boreholes located in this area suggest that much of the succession is clay. In the alluvial aquifers yields up to more than 20 litres per second have been obtained.

According to National Water Resources Master Plan, estimates of recharge have been made by the analysis of flow hydrographs, groundwater level fluctuations, flownets and catchment water balances. The results vary considerably. On the basis of the river hydrographs, the annual recharge is estimated as 15 to 80mm to weathered basement aquifers and 3 to 80mm to alluvial aquifers. In the alluvial aquifers, the recharge will also occur by seepage from the river beds where these are significantly permeable. On the basis of 15mm, the recharge over the country works out to $1,414 \times 10^6 m^3$ per year.

3.3 Water resource utilisation

Consumptive and Non-consumptive Uses of Water

Water resources in Malawi are mainly used in water supply and sanitation, agriculture, irrigation, industry, energy (hydropower), transport (navigation), fisheries, and bio-diversity. The utilization of the resources can be categorised into two groups - consumptive and non-consumptive uses. The consumptive uses include water supply and sanitation, irrigation and industry while the non-consumptive uses include hydropower, fisheries, wildlife, bio-diversity, recreation and tourism. The Malawi government has made some tremendous efforts in developing its water resources and these include the ones shown in Table 6.

Uses	Amount (m³/s)	Remarks
Irrigation	32	Consumptive use
Hydropower	185	Non consumptive use
Water Supply	3	Consumptive
Industrial and others	1	Consumptive
Total	221	

Table 6. Consumptive and Non-consumptive Uses of Water

Domestic Water Uses

The goal of the Sector is to provide clean water and adequate sanitation to the total population in the long run, and to achieve coverage of 84 percent by 2010 in the medium term. Currently, more than 65 percent of the country's population has access to safe drinking water. Access to potable water is higher in the urban than rural areas with 85 percent and 45 percent of the population respectively. Nearly all the residents of the two cities of Blantyre and Lilongwe have access to Municipal water supply.

The Government has over the years invested in water supply. Over 27,000 boreholes, more than 79 rural gravity-fed piped water supply schemes, and 55 municipal and peri-urban water supply schemes have been constructed over the years. These systems altogether provide access to potable water to about 65 percent of the country's population. Boreholes, shallow wells and gravity-fed piped water supply schemes predominantly serve rural communities. The gravity-fed schemes have over 10,000 tap points serving more than 1,200,000 people. The installation of these schemes started way back in 1965. The sources of these schemes are rivers from forest reserves. In the past, the catchment areas of the schemes were un-encroached with good water qualities hence most of these schemes do not have treatment facilities. At present it is only Mpira/Balata scheme which has water treatment facility.

These schemes or water points are designed to provide at least 27 liters per capita per day within a walking distance of less than 500 metres. The provision of these services were originally the responsibility of government, but over the years there has been some involvement of non-governmental organizations (NGOs) and the private sector in the provision of water services especially in the rural areas. The provision of urban water supply services is done by the five Water Boards namely Blantyre Water Board (BWB), Lilongwe Water Board (LWB) and Regional Water Boards (Northern, Southern and Central).

Over the years, Improved Community Water Points (ICWP) have been installed throughout the country. According to the Ministry of Irrigation and Water Development, one is said to be accessible to safe drinking water if a functional ICWP exists within 500 metres of one's home. Recommended maximum number of people using one water point is 250 for borehole and 120 for standpipe.

A big proportion of the country's rural population is without access to potable water. Table 7 shows availability of water supply facilities to rural people in 13 selected districts in the country. The picture is expected to be the same for the rest of the country. Average proportion of the rural population of the 13 districts without access to potable water is 34 percent. The average is expected to be more or less the same for the rest of the country. In other words, the current coverage of potable water supply in the country is about 65 percent which may go as low as 40 percent due to non-functionality of the facilities at any one time. However, Government's target is 84 percent coverage by 2010, according to the Malawi Poverty Reduction Strategy Paper. It can be concluded that this target may not be achieved at all.

The traditional sources of water supply in Malawi are open hand-dug wells usually dug in flood plains or *dambos*, open surface water bodies (rivers, lakes or dams). Access to safe water defined as water piped into the dwelling, public tap, a borehole or protected well, or spring located whether on the premises or less that half a kilometre from the premises) is limited.

District	Rural Population	No. of ICWP	Functionality Ratio (%)	Population Without Access	
				No	%Age
Balaka	277,721	1,588	76	56,156	25
Chikwawa	328,336	1,404	49	179,427	56
Chiradzulu	230,202	1,139	77	51,552	22
Lilongwe	876,476	2,928	63	455,210	52
Machinga	338,899	2,121	59	125,396	37
Mulanje	404,739	2,866	52	146,880	36
Mwanza	128,057	789	72	41,140	32
Mzimba	480,242	4,670	72	104,889	22
Nkhata Bay	138,390	1,229	64	3,668	3
Phalombe	217,729	1,637	46	88,438	41
Salima	219,730	1,148	78	66,541	30
Thyolo	437,361	1,271	83	241,141	55
Zomba	460,538	3,332	61	142,146	37

Source: Malawi Rural Improved Community Water Point Inventory Draft Report, 2004

Table 7. Availability of Water Facilities or ICWP

It is estimated that up to 30 percent of the facilities are out of order at any given time. In total, some six million rural residents are exposed to health risk caused by lack of potable water hence become vulnerable to water borne diseases. Coverage would be higher if fewer systems were non-operative, malfunctioning or dried out as a result of extended drought episodes.

Great variations in service provision occur at district level. Approximately 65 percent of the population of Rumphi district have convenient access to safe water, while less than five percent of the population of Ntchisi and Mwanza districts are adequately served. Levels of access in Ntchisi District would rise if the acceptable distance between dwelling and water point is increased to one kilometre. This, however, is not the case in Mwanza District where only 4.9 percent of the population have access to safe water within one kilometre from a safe facility such as a borehole in rural areas and a public tap in urban.

The rapid urbanization is of increasing concern. The urban growth rate is currently estimated at 6.1 percent nationally, and 10.6 percent for the northern city of Mzuzu. Between 50 percent and 80 percent of urban and semi-urban residents are accommodated in "traditional housing", according to the National Water Resources Master Plan of 1986. These people normally constitute the poorest sector of the urban population. Growth rates of some peri-urban communities around Lilongwe have been estimated to be as high as 15 percent per annum. Population increase could therefore add a further 1.5 to 2.2 million to the number of people without safe water by the year 2010.

Hydropower Generation

Demand for water for hydropower generation continues to be high. Though it is not a consumptive use, large amounts are allocated for power generation in the Shire and in the Northern Region. In the Shire River, hydropower plants of about 200MW generation based on a minimum flow of170 m^3/s were developed after the construction of Kamuzu Barrage at Liwonde, in 1965. However, this design flow was based on the assumption of steady Lake

Integrated Water Resources Management - Key to Sustainable Development and Management
of Water Resources: Case of Malawi

159

Malawi levels of 474.00 m.a.s.l and the existence of a regulating dam at Kholombidzo, some 180km downstream of the lake outlet. This hydropower accounts for the production of more than 98 percent of the total electricity consumed in Malawi. Although a small percentage (3-4 percent) of the country's energy needs, it is the primary source of energy driving the economic and industrial infrastructure and services.

Agriculture

Agriculture is the mainstay of the country's economy. The sector contributes about 36 percent of the Gross Domestic Product (GDP), 87 percent of total employment, supplies more than 65 percent of the manufacturing sector's raw materials, provides 64 percent of the total income of the rural people, and contributes more than 90 percent of the foreign exchange earnings. It is the main livelihood of the majority of rural people, who account for more than 85 percent of the current estimated 12 – 13 million people.

Agriculture is the largest consumer of water in the country. About 70,000 hectares of land in Malawi have been developed for irrigation mostly for sugar and on a small scale for smallholder rice schemes and tobacco estates. Of this, more than 20,000 hectares are being used for commercial farming in the Lower Shire Valley and the lakeshore. Irrigation, demand accounts for about 20 to 25m³/s. However, a further 50,000 hectares are under traditional cultivation. Investment in irrigation has also been on the increase in an effort to boost agricultural production.

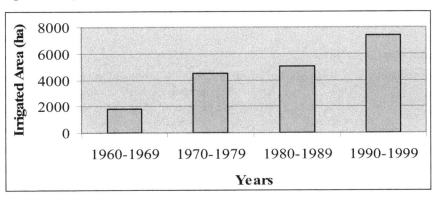

Source: NSoER 2002

Fig. 2. Smallholder Irrigation Scheme

Fisheries

Malawi's water resources have a major role in fisheries development, wildlife and biodiversity promotion and conservation. Major water resources of Lake Malawi and the Shire River remain the major habitant areas for fisheries, wildlife and biodiversity promotion and conservation. The unique relationship between the Upper Shire River and the southern end of the lake is an ideal and very important ecosystem for fish breeding and development. The Shire River also acts as a natural system for disposal of background effluent. There are also a number of investments on tourism and recreational industries in the form of permanent structures and settlements which have been established at the waterfront along the lake shore and the Shire River banks

Transport and Navigation

Lake Malawi and the Shire River play a vital role in the transport system of the country. The Lake provides the cheapest means of transportation through navigation and forms a very important component of the northern corridor transport network between Malawi and Tanzania. It brings about integration of transport systems of Malawi, Mozambique and Tanzania. To promote the water transport, a number of facilities such as ships, boats, harbours and ports have been built. An estimated amount of 180,000 tons of cargo and 300,000 passengers are handled annually.

Tourism and Settlements

The water resources are also important for permanent settlements and for the tourism industry in the country. The Lake Malawi Shire/River system is a particular attraction to tourists and the shores are dotted with settlements and recreational facilities such as hotels, motels, and cottages. Tourism is important in boosting the country's economic development as a source of foreign exchange earnings.

Environment

Environment is recognised as a legitimate water user. National Water Policy has allocated a minimum of eight percent of water to the environment. This is much lower allocation than provisions made by other African countries such as South Africa and Kenya who have allocated 30 percent for environmental flows. There is need to revisit the proportion of water allocated to environment flows.

3.4 Water resource issues

Water resources are increasingly threatened by a number of factors such as prevalent water scarcity, water resources degradation and pollution, over exploitation, and conflict of interest emerging out of lack of integrated water resources development and management, and rapid population growth. These greatly reduce the availability of water for multi-purpose usage.

a. Issues on Water Resource Availability

Water Resources Scarcity

Most rivers run dry by July, with the exception of those flowing from high altitude-rainfall areas of Nyika and Viphya plateaux, the Kirk Mountain Ranges, Zomba Plateau and Mulanje Mountain. This situation of unreliable dry season flows has been exacerbated by deforestation and land use malpractices.

These have particularly occurred in headwaters, escarpment and mountainous catchment areas, which are in normal circumstances, supposed to exist as protected land. This existing situation has led to the flashy nature of surface runoff without significant recharge of the aquifers, thereby greatly diminishing the base flows in the water resource basins.

Water Resources Degradation

The degradation of catchment areas including marginal lands due to population pressure, deforestation and poor agricultural practices have facilitated the development of soil erosion leading to serious sedimentation or siltation problems. The rapid increase in population

Integrated Water Resources Management - Key to Sustainable Development and Management
of Water Resources: Case of Malawi

161

densities and the growth of industries have inevitably worsened the already existing problems on the demand for land and all other natural resources. This has led to the encroachment of marginal and water catchment areas. Poor land management practices with resultant erosion of the soil have led to siltation of rivers and reservoirs causing serious water quality problems downstream and inundation of river channels as a result of increased surface runoff. Inundation of the river channels causes flooding of rivers resulting into the destruction of crops, people's houses and property as well as loss of life. This is mostly occurring in low-lying areas of the lake shore plains such as Karonga, Salima-Nkhotakota, Bwanje Valley, and Lower Shire Valley.

The quality of surface water resources has substantially deteriorated due to several factors that include; inappropriate land use practices, improper and in some instances unwarranted usage of heavy agrochemical and unchecked disposal of domestic and industrial wastes and allied effluent matter. The quality of groundwater is however generally good, although isolated and sporadic occurrences of saline ground intrusion are encountered. In some areas the utilisation of ground water for water supply has been limited by the presence of high contents of parameters such as iron, fluoride, sulphate and nitrate.

A number of rural water supply schemes which the government had constructed in the early 1970's with minimal treatment facilities and relying heavily on abstracting water from protected catchments (like forest and wildlife reserves) have recently been heavily contaminated by human faecal coliforms as a result of human encroachment into these protected areas.

Non-designated damping sites operated in line with unplanned and poor technological operational concepts that do not conform to stipulated health or any scientifically developed standards are known to lead to pollution and other detrimental effects on the water resources of the country.

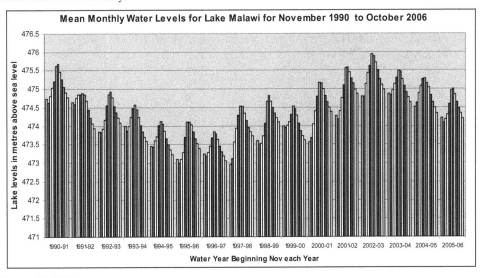

Source: MoIWD

Fig. 3. Lake Malawi Trends in 1990/1991

Drought and flooding are recurrent problems in Malawi. The impact of climate change and variability strongly influences the occurrence and distribution of floods and droughts. The late start of the 2005/2006 rainfall season and inadequate rainfall during the season resulted in dwindling of water resources. This was clearly evident in surface water resources as many rivers have had lower flows in the past water year. Even the lake levels have experienced a significant drop. For example, the mean lake levels for October 2006 was 474.19 m.a.s.l. compared to 474.35 m.a.s.l. for October, 2005 and 474.65 m.a.s.l. for October 2004. The declining trend of lake levels from the peak of 2002/2003 shown in the Figure 3 below may be a reflection of the upland catchment rainfall impact.

These issues are contributing to the imbalanced situation of the water resources and show the great need to check and arrest the situation if the set development goals and objectives are to be achieved. It also calls for a holistic approach in the development and management of water.

b. Issues on Water Resource Utilisation

The available water supply systems are most vulnerable to the effects of droughts and unreliable dry season flows. This is so because very few systems have reservoir storage facilities to act as back-up to the supply system which proves to be of strategic importance during low flow seasons or no-flow periods. That is to say most of the developed systems rely on run-of-the-river water supply schemes, which are heavily susceptible to the effects of hydrological droughts and seasonal fluctuations.

The water delivery services in the country, including those relying on boreholes and wells are also adversely affected by poor design values coupled with inappropriate operation and maintenance mechanisms.

The population with access to potable water drops from 65 percent to levels as low as 40 percent at anyone time due to problems of water scarcity and those affecting operation and maintenance of the delivery system. The remaining population of 35 percent access water from unprotected sources resulting in the high prevalence and upsurge of water borne or water related diseases. The trend overtime however, show a slight increase of people accessing potable water (Figure 4).

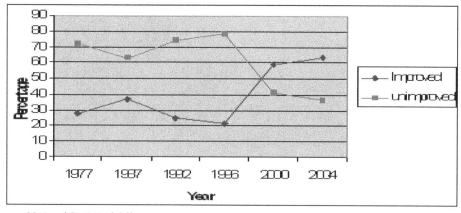

Source: National Statistical Office

Fig. 4. Safe Water Access Trends for Malawi

Even though more than 99 percent of the country's power is hydro-generated, the high tariffs still cut off majority of the population from accessing electricity. This has forced most people to depend on fuel wood thus causing uncontrolled deforestation. To satisfy these demands, there is need for a multi-sectoral administrative coordination, and enhancement of political, legal and economic co-operation initiatives that can assist in the regulation of hydrological regimes not only for Lake Malawi and the Shire River, within acceptable limits, but also for all water resources of the country as a whole.

4. Water resources management framework

Water resources management basically involves the monitoring and assessment of water quality and quantity, the development and protection of water resources, the provision of water services and ensuring that water laws are strictly adhered to by all users. In other words, water resources management may be defined as man's control over water as it passes through its natural cycle, with balanced attention to maximising economic, social and environmental benefits. The goal of sustainable IWRM is, therefore, to conserve water resources in both quality and quantity for the benefit of the present and future generations.

Pressure on water is increasing due to increase in population. Land use is intensifying causing increased land degradation and eventually degradation of the water resources. This causes an increase in demand for water and a range of potential and actual threats to the quality and quantities of water available. Also water supply and water-borne sanitation services are inadequate to meet the needs of some communities. This calls for action from water sector and water related sector organisations as well as the user communities in the proper planning, development and utilisation of water resources to achieve the maximum benefits while at the same time ensuring their sustainability.

4.1 Water policy and legislation

A number of water resources management policies and legislations have been enacted in Malawi. The policies and legislation have been regulatory, to promote conservation, equitable allocation and protection of the resources against pollution, over-exploitation, and physical degradation to establish water supply and sanitation delivery services or other water dependent services. The Water Resources Act (1969) and its subsequent amendment and subsidiary Water Resources (Pollution Control) Regulations provide the main regulatory framework for water resources management. On the other hand, Water Works Act (1995) is the main authority for establishing water supply and water borne sanitation delivery services.

The government introduced the 1994 Water Resources Management Policy and Strategies (WRMPS). This policy and its strategies, however, have been reviewed by the 1999 and 2005 Water Resources Management Policy and Strategies funded under the Natural Resources Management and Support (NATURE) programme. The introduction of the National Environmental Policy and the call for harmonisation of natural resources management policy, legislation and institutional roles, warranted the review. This provides an enabling environment that is conducive for sustainable and integrated water resources development and management.

The policies and legislation of other sectors benefiting and affecting the water field need to be in harmony with those drawn for water resources management. One of such important

Acts is the Environmental Management Act, which serves as an umbrella legislation with its aim focused on enforcing sustainable development of the natural resources. It is also important for the national water resources legislation to be in harmony with those of the neighbouring countries in the region. The signed Protocol on Shared Watercourse for the SADC region (2000) provides a framework for harmonisation of policy and legislation among the SADC and Zambezi River Basin member states.

The revision of the policy and legislation on water resources management entailed the revisiting of institutional roles of the Ministry of Irrigation and Water Development and harmonising them with other government departments and agencies involved in the water sector and other organisations whose undertakings are in any form linked to water resources management and development. This will ensure that policy and legislative reforms are documented and implemented without duplication or dormancy in any essential water environmental resources management services.

4.2 Existing framework for water resources management

The water sector in Malawi comprises several levels of responsibility that range from national policy formulation through the administrative and management units down to service providers in the construction, operation and maintenance of water supply, water borne sanitation and any other water based or related fields. These levels of responsibility are assigned to different government institutions and parastatal organisations of utmost importance recognising how the MoIWD has developed over the years to lead the sector amidst many challenges to where it stands today.

4.3 Evolution of the water and sanitation sector

Water as a distinct sector in Malawi is relatively new and still developing. The process to develop a distinct sector in Malawi, like many other countries in the World, has evolved over time in response to the growing need for better management and development of water resources to meet the growing demand.

During the colonial and federal administrations there was no real effort to develop water resources for social-economic development or to meet the needs of the wider indigenous population even though 750 small dams were developed during that time. Therefore, few systems were installed in urban centres mainly to service the expatriate community. Different water related functions were therefore scattered in different Government units only as incidental activities in the line functions of those units.

After gaining independence in 1964, the new Malawi Government continued with the same institutional framework, hence despite the increased exploitation of water resources mainly for agriculture development, there was no sectoral consolidation of water related development. Similarly, these were scattered in many Government units and uncoordinated, for example, Water Resources Management and the branch of Irrigation were in the Ministry of Agriculture and Natural Resources. Ground Water Division (Boreholes) was in the Geological Survey, while Rural Water Supply was under Community Services. Some urban water supplies were functions of city/town councils under Local Government, and District Water Supplies were under the Department of Public Works.

Integrated Water Resources Management - Key to Sustainable Development and Management
of Water Resources: Case of Malawi
165

This disintegrated approach to water related developments greatly retarded the proper management and development of water resources in Malawi due to lack of direction, coherent planning and efficient use of the human and other material resources. The outcome of this was rapid deterioration of the country's water resources and little coverage in terms of provisions of safe water to both urban and rural populations between 1964 and 1980.

In order to raise the profile of Water and Sanitation as a key ingredient of socio-economic development in the world, the United Nation introduced the water supply and sanitation decade in the 1980's. As a requirement for Malawi to benefit from this water development initiative, a Water Division was created in 1979 under the Department of Lands, Valuation and Water in the Office of the President and Cabinet. This was the first step towards the creation of distinct sector, and it brought together the hitherto scattered units of water resources, groundwater, rural water supplies, district/town water supplies, water quality and irrigation.

The profile of the new water sector was enhanced in September 1984 when the Water Division was separated from the Department of Lands and Evaluation and a fully fledged Water Department was created under the Ministry of Works. In September 1994 the Department was given full ministerial status as the Ministry of Irrigation and Water Development, and then after a functional review in 1997, the Ministry of Water Development was created. This led to clear separation of institutional roles among water boards, regional water boards, and how the Ministry would interact with various players in water resources development and management. In 2004, the Department of Irrigation was separated from the Ministry of Agriculture and Food security and rejoined the Ministry of Water Development to become what is today called the Ministry of Irrigation and water development.

5. Water resources management key challenges, opportunities & risks

A major challenge facing the IWRM in the country and world over is the understanding of IWRM concepts by the decision makers. In most cases the IWRM concepts are poorly understood in spite of the recent trends by the United Nations in recognition of the need to adopt IWRM principles for sustainable water resources management.

Important hydrological and ecological services are considered to be of marginal value and it is therefore not surprising that environmental issues continue to receive less attention from most central governments and its protection is seen by decision makers as a "green" issue promoted largely by external interest group.

Water is a necessary input for any productive activities of agriculture, forestry, industry, mining, livestock development, energy production, tourism, wildlife conservation and domestic water supply among others. Therefore, the effective integrated management of water resources and sustainable utilisation is a pre-requisite for sustaining all forms of life and fostering overall socio-economic development in Malawi.

Water availability varies considerably within Malawi. Overall it is a scarce resource, which is vulnerable to global factors such as climate change, and to regional constraints imposed by the management of trans-boundary waters. Water is also vulnerable to local and national factors such as the growing demands of urban and rural populations, increasing sectoral demands, greater competition and potential for conflict over water, worsening water pollution, land and catchment degradation, destruction and encroachment on aquatic ecosystems, and proliferation of invasive weeds. Increasingly, environmental degradation from unsustainable land and water use patterns and other anthropogenic factors is undermining and threatening the sustainability of the water resource base itself. These challenges are likely to further exacerbate the water scarcity in the country if they remain unchecked. The challenges have been summarised as follows:

5.1 Water scarcity and droughts

There are inadequate water resources to meet demand due to increased seasonal variability in run-off, increases in population and demand for industrial production, and irrigation requirement. This problem has basically arisen as a result of droughts and unreliable dry season flows. 1991/92 – 1994/95 droughts showed the vulnerability of water supply schemes. Most of them rely on the direct run of the river which. Low availability of seasonal run-off greatly affects the hydrological demand for domestic needs, industrial production, and irrigation requirements.

There are frequent occurrences of droughts, which are initiated by the EI Nino and Southern Oscillation (ENSO) phenomena. These have resulted in inadequate amounts of rainfall received, hence declining amounts of both surface and groundwater resources.

5.2 Water resources degradation

Pollution of surface and groundwater resources is a growing problem and is making water resources unavailable for use without expensive pre-treatment. Pollution is increasing from point source such as effluent discharges of industrial and domestic waste, and from non-point sources such as solid waste, silts, and agrochemicals. The following are some major causes of water resources degradation in Malawi:

a. Effluents Disposal

The disposal of effluents puts a demand on the resources either by reducing the quality of the sources of raw water, or by requiring the regulation of stream flow to provide greater dilution. At present, the demand exists only in the first category, and mostly in the urban areas, such as Blantyre, Lilongwe, Mzuzu and Zomba. Future demand is likely to increase rapidly with the growth of industrialisation and centralisation of population in the urban areas. However, the picture is far from clear and information is urgently needed. Disposal of effluents directly into Lake Malawi is especially an important issue, bringing into light the conflicting requirements of industry, fisheries and tourism.

b. Sedimentation (Siltation)

Great pressure on the land resources resulting in soil erosion and deforestation has been experienced in Malawi due to rapid population growth. Silt loads (sedimentation) in surface runoff from soil erosion lead to significant problems in down stream water quality, including increased suspended solids and turbidity, resulting in high water treatment costs and water flow problems. Since 90 percent of the people reside in rural areas and depend on rivers for water supply, chances of drinking unclean water are very high.

Integrated Water Resources Management - Key to Sustainable Development and Management
of Water Resources: Case of Malawi

167

As indicated in case studies carried out, catchments with high deforestation rates have faced increased levels of flow discharge, which in turn have led to increased levels of turbidity and solids. High sediments loads in the rivers bring about siltation of rivers (when the gradient is low) and of water reservoirs. The silted river courses and of water reservoirs tend to have reduced capacities so that when it rains the banks may overflow, causing flooding at times; or the water erodes the bank (in order to accommodate the increased volume of run-off). The intake point for Nkula Hydro-electrical power reservoir, for instance, is frequently dredged for this reason. Sedimentation may also affect fish resources.

c. Chemical Contamination

With the reduction in fertility of agricultural land, use of agrochemicals such as fertilisers, pesticides for increased productivity becomes inevitable. However, run-off of these chemicals into water bodies can cause serious harm to human health, livestock, fish, and other aquatic environment. For example, these chemicals may result in eutrophication of the water bodies and an increase in water plants growth. This threatens fish resources growth and reproduction.

d. Catchment area encroachment

Encroachment of protected catchments, through deforestation, human settlement and cultivation of marginal lands is an issue of major concern in Malawi today. This nature of pressure exerted on the water resources brings about declining base flows, deterioration of water quality, and reduced groundwater recharged rates, increased turbidity of water in river and reservoirs and increased incidences of flood disasters.

5.3 Institutional and legal constraints

The Government's policy is to provide clean potable water to all people so as to reduce the incidence of water borne diseases and reduce the time devoted by individual to water collection. The Water Resources Act and other Acts that deal with the use of water for different purposes have been found to be inadequate in that they conflicts each other instead of complimenting one another in issues of water resources management. They also lack punitive measures against those who cause substantial water pollution by discharging effluent. The proposed implementation of institutional arrangement or other strategies mentioned in the current policy may take longer time to implement the strategies or the targets. The weakness and strength of these institutions should have been translated into policies to ensure that they have the capacity not only to implement the National Water Resources Management Policies (NWRMP), but also integrated management.

Within the new water policy and strategies, there is an increased effort on decentralisation and ensuring resources are owned and managed at a local level. The roles of the various bodies, including Village Development Committees (VDCs), area development committee (ADCs), District Coordination Teams (DCTs), district assemblies (DAs) and Catchment Management Authorities, require careful analysis to ensure co-ordinated execution. It is important to consider the appropriate entry points for organisations into rural communities to avoid duplication of services in a given area.

5.4 Inadequate capacity to carry implement IWRM

Capacity building should be guided by a clear understanding of institutional roles and responsibilities in the context of decentralisation. A broad view of capacity building is required, including issues of management, human resources, skill development, organisational development, training and the mobilisation of financial resources.

The changing roles of MoIWD require new skills and competence in planning, monitoring, evaluation and regulation of water supply and sanitation services in line with the Decentralization Policy. Capacity building must be tailored to allow the officers to take on these new roles and perform them effectively. It is, however, important to identify and analyse existing capacity gaps and tailor the capacity programme to fill these gaps.

For effective implementation, sustainability and operation and maintenance of water and sanitation projects, there is need for capacity at grass root level. The basics of CBM have become institutionalized in many programmes and projects and this is having a positive impact on the maintenance of water points within communities. There is, however, need for further development to ensure the long term sustainability by improving the spare parts supply chain and the mechanism for carrying out repairs beyond the capacity of the communities and increasing the range of technology choices offered to communities.

Capacity building is not just a role for the Government. The challenge is to see how all the stakeholders can contribute to building capacity within the sector.

5.5 Standards, procedures and specifications

The provision of water and sanitation services is performed by a number of organisations and institutions. This has led to a proliferation of approaches and procedures and a lack of standardisation. Technical specifications for water and sanitation facilities are very variable and this leads to cost variations. Sometimes the bases for these cost variations are not obvious, other than the fact that the different procedures may incur different expenditures. The challenge is to see how the sector coordinates these various approaches and implement them in IWRM.

5.6 Disaster management (floods and droughts)

Currently, Malawi has a reactive approach to flood and drought disaster management. In years of excessive rains, flood disaster management is prevalent, whilst the management of food and water shortages follows the years of droughts. The country is aware of areas that are prone to floods and droughts, but as yet no mitigation measures have been instituted. Limited efforts have been made to utilise forecasting models for flood warning and early drought monitoring, but there is only very limited infrastructure activity.

In 'normal' years the water resources of Malawi are mostly seasonal causing floods during the rainy season and droughts in the dry season, yet per capita availability of water shows ample water resources for every citizen. Water resources will have to be harnessed through water harvesting and catchment rehabilitation and management.

Properly managed catchments will improve water retention on the catchment to enhance aquifer recharge and improved dry season stream flow. The challenge is to implement an integrated flood management initiative that aims at ensuring an end-to-end process of flood

Integrated Water Resources Management - Key to Sustainable Development and Management
of Water Resources: Case of Malawi

169

management, put in place in a balanced manner, duly considering prevention and mitigation measures and the positive and negative impacts of floods. The development of the resultant national integrated flood management and drought management strategies is definitely another challenge

5.7 Shared water resources

Management of Lake Malawi, Shire -Ruo River system, Lake Chilwa, Lake Chiuta Songwe River catchments shall required cooperation between riparian countries of Tanzania and Mozambique. There is undoubtedly a challenge in luring political support to the management of the transboundary water courses. A mechanism is required to create a flat form for consultation and dialogue among affected nations.

The existence of legal instruments of shared water resources provides a favourable environment for contact and dialogue among riparian states in order to promote integrated water resources management.

5.8 HIV/AIDS

The number of deaths and the extent of human suffering caused by AIDS have assumed an unimagined order of magnitude. The consequences are now so pervasive that they constitute a substantial challenge to social and economic development in Malawi. HIV is no longer simply a public health issue but a major concern to sustainable integrated water resources development and management. It cuts across agencies, disciplines and national boundaries.

6. Summary

Malawi is water stressed country with total renewable water resources per capita of 1,400 m³. With such a low per capita, Malawi is worse of than Botswana and Namibia, countries which have large areas of desert. Within the SADC, Malawi is the second country with low per renewable water per capita, after South Africa.

Per capita of renewable water available will decline further over time due to rapidly growing population of 2.8 percent, climate change/climate vulnerability and water quality degradation (due to poor agricultural practices, poor waste management, deforestation and forest degradation). In spite of the low per capita adoption of WDM, strategies has remained low especially in agriculture sector most of the water is used.

Lake Malawi-Shire River water system, which is a strategic water resource for hydro power generation, irrigation, navigation and fisheries, is a vulnerable resource because about 53 percent of the water comes from the catchment in Tanzania. Any major water development activities within the catchment would have serious consequences of the economy of Malawi. Efforts to manage water resources more efficiently have been hampered by inadequate capacity for IWRM, unharmonised policies and laws, inadequate catchment management practices and poor coordination among stakeholders.

Water stress status of Malawi is a serious threat to development of the country and has the potential to reverse development already achieved by the nation. Water shortage will seriously affect efforts of Malawi Government to achieve Growth and Development goals set out in MGDS because MGDS is about using more and more water for farming, tourism, industries, navigation, electric power generation and other economic activities.

The declining water situation will now become the major limiting factor towards development of the country. Water allocation among competing potential users will become critical and trade offs will have to be made in order to ensure that the scarce water resource is used in an activities that will result in maximizing benefits for the nation.

The trans-boundary nature of the water resource also mean that Malawi will need to develop very close dialogue/consultations with the neighbouring countries in order to ensure that water development in respective countries does not negatively affect the development agenda of neighbouring countries. While joint commission of cooperation exists with Mozambique, Malawi needs similar consultation mechanism with Tanzania and Zambia. Issues of common interest will be both water resources management plans in neighbouring countries but also Malawi will need to start dialogue on cross border water transfer potential from neighbouring countries to Malawi.

IWRM/WE offers an approach that can enable Malawi effectively address national and international water resource challenges by promoting integrated management of natural resources and promoting consultations of various stakeholders in water and water-related fields.

7. References

Carl Bro International, 1995. *Sanitation Master Plan for the City of Blantyre*: Existing Sanitation Situation Volume III. Government of Malawi.
Chipofya, V., Kainja, S., Bota, S., 2009. *Policy Harmonization and Collaboration amongst institutions – a strategy towards sustainable development, management and utilization of water resources: case of Malawi*. Desalination, Elsevier. 248, Issues 1-3, 678-683.
Malawi Government, 1969. *Water Resources Act*, Cap 73:03, Laws of Malawi, Government of Malawi, Zomba.
Malawi Government, 1995. *Waterworks Act*, No. 17, Laws of Malawi, Government of Malawi, Lilongwe.
Malawi Government, 1995. *Malawi's Vision 2020*. Government of Malawi, Lilongwe, Malawi.
Malawi Government (2003). Strengthening of Water resources Boards Report, Annex 6
Malawi Government, 2005. *National Water Policy*. Ministry of Irrigation and Water Development, Government of Malawi, Lilongwe.
Malawi Government (2006). *Ministry of Irrigation and Water Development National Strategic Plan* (2006 – 2011).
Ministry of Works & Supplies/UNDP (1986). *National Water Resources Master Plan*, Report and Appendices: Surface Water Resources General.
Malawi Government, 2007. *Integrated Water Resources Management and Water Efficiency Plan*. Ministry of Irrigation and Water Development, Government of Malawi, Lilongwe, Malawi.
Malawi Government, 2008. *National Sanitation Policy*. Ministry of Irrigation and Water Development, Government of Malawi, Lilongwe.
Malawi Government, 2009. *Population and Housing Census*. National Statistical Office, Government of Malawi, Zomba www.un.org/millenniumgoals/

8

Indicators of Traditional Ecological Knowledge and Use of Plant Diversity for Sustainable Development

Edmundo García-Moya[1], Columba Monroy-Ortiz[2],
Angélica Romero-Manzanares[1] and Rafael Monroy[2]
[1]Colegio de Postgraduados,
[2]Universidad Autónoma del Estado de Morelos
México

1. Introduction

Mexico is a country biologically and culturally diverse, historically linked to peasant and indigenous people who derived livelihood from nature and contributing to generate the traditional ecological knowledge, which leads to sustainable resource management adapted to the availability, needs, and options. Internationally, it has been suggested the involvement of indigenous and local communities to generate plans of social development, biodiversity conservation, environmental impact assessments, enact laws, and for the generation of criteria and useful indicators in the evaluation and monitoring of the utilization of natural resources. The goal of sustainable development is to meet human needs and aspirations through the development of human progress towards protecting the future (ONU, 1992).

Sustainable development indicators used by the international community tend to be based on information available at the official statistics of countries, however, there are elements of each nation that are essential for survival and social development, and are excluded from official assessments. In Mexico, the contribution of indigenous and local communities to national development is not quantified, although these communities protect much of the biological and cultural diversity of the country (Boege, 2009). Biodiversity is essential for the survival of a social sector permanently subjected to exploitation and marginalization, which represents approximately 80% of the population nationwide (Boltvinik & Hernández-Laos, 2000).

Together with the global social development, indicators are required at national level to meet the particularities of countries like Mexico, which account with great cultural and biological heritage, in which social participation has been limited for educational, legal, or lack of access to timely information. Even more difficult, is the acknowledgement of the traditional ecological knowledge and the vision held by people of indigenous and local communities. The generation of environmental indicators and use of biodiversity based on traditional ecological knowledge is an opportunity to learn about cultural diversity, to describe the current health of the ecosystem, and monitoring the changes of this kind of knowledge, since as far as this kind of knowledge maintains its functionality it might meet the needs of the stakeholders. The contribution of the indicators could transcend into the evaluation and follow up of the actions undertaken to achieve social development. This proposal is viable in Mexico because there are numerous research papers on Ethnobiology providing information about the knowledge underlying the utilization of biodiversity. Also it is relevant as long as, it has been probed the economic importance of forest for the people who take advantage of them and utilizing the traditional ecological knowledge.

2. Background

Traditional ecological knowledge has kept for some time and the value of this cumulative and dynamic process, practices, experiences, adaptation to local circumstances, holistic and useful, has taken up to improve the capacity of societies to manage natural resources, especially in changing and uncertain conditions. The value of this knowledge applies to various disciplines including business, such as the search for active ingredients of plants which increases the efficiency of the process 400 times (Reyes-García, 2009). However, according to the author, it is clear that the over-exploitation of common pool resources (Ostrom et al., 1999) is a problem linked to others that they occur in chained form, for example, over-exploitation is the result of open access to resources leading to the collapse of production, erosion of traditional systems of resource management and lost of traditional ecological knowledge, according to the incorporation of indigenous or rural communities into the market economy or school education.

Taxonomic groups as indicators of diversity have become useful in the characterization of certain cases for the identification and resources utilization, however, they reflect snapshot conditions of a situation (Halffter & Moreno, 2005), more than development of spatial and temporary processes. The importance of sustainability in the loss of resources is increasingly, yet, the common contempt for the indigenous communities as sites of consensus when it comes to resource management (Natcher & Hickey, 2002), biased decisions unilaterally in social development programs. Persha et al. (2011) shows that forests tend to be sustainable when users have the right to participate in government programs, formally recognized.

2.1 Traditional ecological knowledge

Traditional ecological knowledge is rooted in the population, emerges as a complex of components such as knowledge, beliefs and practices (Berkes et al., 2000), inherited through generations, and whose significance is denoted by maximizing the diversity of livelihood options that evolve through processes adaptive, encourage diversity to face environmental, social, economic or political, with minimal risk (Toledo et al., 2003). There are synergies of the direct and indirect influences affecting the components of traditional ecological knowledge (Millennium Ecosystem Assessment, 2005; Challenger & Dirzo 2009). Indirect factors include the development models that foster inequality and poverty, discriminatory government policies, and institutions and local knowledge holders, limited participation of citizens and the local population during the implementation and evaluation of development models; in the meantime the direct factors include a lack of consideration for cultural diversity, availability of goods and services more accessible and affordable, health systems predominantly based on allopathic medicine, landscape fragmentation, environmental degradation, and cultural erosion among others.

Traditional ecological knowledge is an expression of cultural diversity which has been instrumental in the conservation of resources and meeting basic social needs, supports the management of natural resources by peasant strategy of appropriation of nature, ability to characterize by maintaining high levels of diversity, to promote the resilience of ecosystems which have been sustainable (Toledo et al., 2003).

Five of the dimensions that conform the peasant strategy to maximize social commodities are: use of families, genera and species, use as many native species, maximizing the number of uses assigned to the species, integral utilization of species, and maximizing use of available species in the wild. Consequently, it is necessary to understand and respect the culture of communities, while protecting the traditional knowledge systems.

Culture is key to achieving sustainable development (UNESCO, 2010). The information generated by ethnobiology, scientific discipline that investigates and systematized traditional ecological knowledge, provides information on the use of the species. The information obtained is adapted to function as indicators of the state of resources and production potential for matters concerning the management of diversity. The Commission on Sustainable Development (CSD, 2000) proposes various indicators related to social, economic, environmental and institutional as well as to specify information that is also comparable between different entities.

So far there are laws, regulations, standards and indicators related to biodiversity, for example, the National Environmental Indicators, which includes among others: the production of timber resources, the number of species found at risk or protected natural areas. However, lacks of instruments to consider the interaction of biological diversity with cultural diversity. One argument in favor of the creation of such legal instruments and environmental planning is the overlapping of priority areas based on their biological characteristics, the territories of indigenous peoples and communities in Mexico. On the other hand, the participation of traditional ecological knowledge in decision making is political and social limited for various reasons, including the ideology held by people of indigenous and local communities.

Actually there are technical and financial difficulties to obtain information on the interaction between biological and cultural diversity throughout the country. Meanwhile numerous research papers have documented traditional ecological knowledge that underpins the use and benefits of diversity in rural communities, towns and biogeographic regions. Therefore, these works have the potential to clarify or modify laws, regulations and rules in light of the conditions of biological diversity and cultural knowledge available. One area of opportunity for the incorporation of traditional ecological knowledge in planning instruments is the generation of Environmental Indicators and Use of Biodiversity, inclusive of the needs and characteristics of the cultures involved, so that, having this type of indicators is possible: a) a description of current state and the transformation of traditional ecological knowledge on natural resources, and b) knowledge of the biodiversity that is being used for self-support (needs of the community without marketing purposes). This type of information accumulated historically and of intangible value, goes unnoticed in the statistics on domestic natural resources, like medicinal plants.

2.2 Indicators

One way to enhance the contribution of traditional ecological knowledge for subsistence, conservation and sustainable use of biodiversity, is the generation of indicators. There are instruments to assist decision making for the planning of sustainable development (UN, 2009). The indicators vary according to the definition of purpose, are descriptive, based on components, objects or processes, normative (prescriptive and assessed for changes in condition or efficiency of management), or hybrids (descriptive and normative) (Heink & Kowarik, 2010). In general, indicators should be useful to diagnose an environmental situation, to evaluate the condition or state of the environment or to provide early signals of change. Therefore, should include attributes that would inform the structure, function and composition of ecological systems (Dale & Beyeler, 2001). The scale at which indicators should be generated varies from small local to larger scales, inclusive of provinces, regions and countries. Indicators should also consider an interaction of both development elements: environmental and social, and should link culture with nature.

Internationally there are successful experiences around the participatory generation of indicators in the conservation, management and evaluation of action research for agriculture, the quality of water bodies, local coastal management and environmental degradation. As an example, Munis de Medeiros et al. (2011) used five indicators picking patterns and perception, to estimate the pressure of use, and three categories of impact, due to timber harvest in northeastern Brazil. The indicators are related to the volume of wood consumption, indices of diversity and equity. The conclusion points out that the use of wood energy facilitates the use of greater importance and recommends to specify which species are the highest quality fuel, taking into account the preference of consumers, ecological information in plants, forest seral stage, and post-disturbance regenerative capacity.

It should also be considered under the concept of perception, and what importance do people give to plants or plant uses. In ethnobotany, number of mentions of a plant by the respondents, is an indicator of the intensity of use of a species (Muthcnick & McCarthy, 1997; Begossi et al., 2002). We believe that linking culture with nature through traditional ecological knowledge could be useful for decision-making in the development of policies, plans and development programs.

2.3 Sustainability

In the case of the sustainability of resources, it is important to consider the role of local community initiatives, as in the case of Belize where maintenance of wildlife depends largely on the protection of land through patrolling and systematic evaluation by local users who own the land and operators of resources, including protected areas by the government (Horwich et al., 2011). Sustainability indicators are used in different ways, also in terms of objectives, some of which are only environmental, social or financial, while some are multi-scale, hierarchical, and in other cases, are specific inputs for political reasons aimed at sustainable practices, viewed through systematic observation of changes in holdings (Caceres, 2009). The Framework for the Evaluation of Systems Management of natural resources through Sustainability Indicators (MESMIS) has excelled in Latin America and especially Mexico, as it proposes a systematic framework, participatory, interdisciplinary and flexible approach to assessing the sustainability of agricultural systems. Some of the rates quoted by the author are

EI: ecological index based on physiognomy, erosion and vegetation covers; SI: socioeconomic index based on family income and food security; and LSI: land sustainability index supported by the articulation factor of socioeconomic index plus the ecological index based on specific variables. It has been suggested that achieving a positive impact on biodiversity must be considered as the aggregation of variables and temporal scales of study. To Heink & Kowarik (2010), sustainability is a complex indicandum because it is additive composed of a set of indicators, as outlined in the previous reference in the operation of the term which depends on its original conceptualization. They reiterated that sustainability is a multidimensional indicandum including environmental compatibility, social acceptability, justice and economic development.

Natcher & Hickey (2002) have included in a small area near Alberta, systematic recording of changes in management plans based on objectives, criteria, indicators and actions. In the case of forest management, they consider that the native community should be helped to access to land and resources, protection of the areas identified by members of the community, as important, meaningful historical, biological, cultural, and protection rights to perform various activities in the forest, promote economic opportunities for members of the community, and enabling their participation in the decision-making processes.

3. Contributions based on traditional ecological knowledge and indicators

Due to the rapid transformation of the natural environment in the state of Morelos, Mexico, we need indicators based on the potential of biological and cultural diversity, to generate sustainable development decisions. In this chapter we contribute with some useful indicators for assessing the status of traditional ecological knowledge and the use of biodiversity by rural communities of the state of Morelos (1.7×10^6 inhabitants in an area of 4893 km^2), Mexico. This region occupies 0.2% of the area of Mexico and has tropical deciduous forest in the warm zone, where we have research on the traditional ecological knowledge generated by the local people. The indicators studied were based on the strategies for local use. We pretend that the cultural value and the intangible and implicit traditional ecological knowledge should be known, capitalize and transform in cultural, ecological, economic, and social value useful for sustainable social development, and also serve as an instrument to regulate public policy of utilization of resources.

Should be mentioned as background that we have worked with local traditional knowledge for sustainable non-timber forest species, assuming that traditional knowledge related to using a group of plant species is widely known among local people. We got ethnobiological information from 1979 up to now, based on structured interviews and participatory workshops, to learn about traditional ecological knowledge and non-timber forest species of local importance for use and conservation purposes. Also, we have studied the multiple and comprehensive utilization of resources for firewood (Monroy & Monroy-Ortiz, 2003), charcoal and orchards. As a result, 24 Ethnobiological reports concluded up to 2001 which were systematized (Monroy-Ortiz & Monroy, 2006). Likewise, we used bibliographic information to elaborate databases for medicinal plants (Monroy-Ortiz & Castillo, 2007). For the analysis of local knowledge and indicators using descriptive and multivariate statistics, fieldwork as check up applying the scientific method coupled with demographic assessment of floristic resources and uses (Monroy-Ortiz et al., 2009). In the next lines we mention the wealth of plants used some of the general traits of traditional ecological knowledge registered; we discussed briefly the importance of medicinal plants and also the form of supply of firewood.

3.1 Richness of medicinal plants used

We recorded 581 species belonging to 402 families used and 130 botanical families. The absolute values of the services registered (Figure 1) include different plant parts that meet the various needs of society, so that traditional ecological knowledge held by people in the communities studied, allow them to obtain a wide range of provision, regulation, supporting and cultural services.

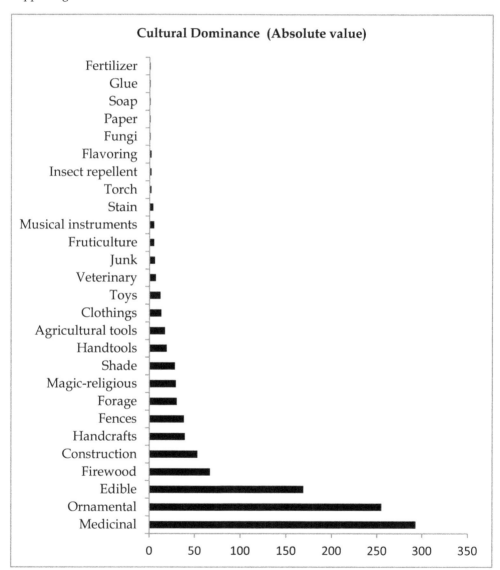

Fig. 1. Provision, regulation and cultural services of the tropical deciduous forests of the State of Morelos.

Highlights the medicinal, ornamental and edible uses, the first two are essential for survival and reproduction of society (Gispert & Rodriguez, 1998). Medicinal and edible uses are closely related to each other by facilitating the free flow of species. The ornamental use is related to the favorable climate for cultivation, making it economically productive. It is interesting that in the group of ornamental plants there is a notable tendency toward homogenization of diversity and the replacement of native flora by introduced species such as *Ficus benjamina* L., which replaces to the native urban woodland of *Plumeria rubra* L. or *Tabebuia rosea* (Bertol.) DC., two trees noted for its beautiful flowers. It should be mention that there is a transition in the use of an ornamental species for medicinal including exotic species that are empirically evaluated by their medicinal potential (Bennett & Prance, 2000). This is the case of *Ricinus communis* L. a plant native to Africa that is used in Morelos to elaborate a poultice that is applied on the feet and abdomen to reduce fever. The transition from the ornamental to medicinal demonstrates the importance of underestimating the types of use considered non-dominant and expendable.

We think it is necessary to examine whether the transition of uses due to losing a plant resource or traditional knowledge associated with it, it would encourage a multiplier effect of cultural erosion.

3.2 Importance of using medicinal plants in Morelos State

A total of 821 species of medicinal plants are used to cure diseases of 22 kinds of systems and apparatus (Monroy-Ortiz & Castillo 2007). Just for the digestive system there are 453 plant species, 54% of the 842 species. For trauma treatments the people use 224 species (27%). The prevalence of digestive illness is probably related to poverty and extreme poverty in which they live 80% of Mexicans. Other illnesses treated with plants, according to traditional knowledge, for the state of Morelos are shown in Figure 2.

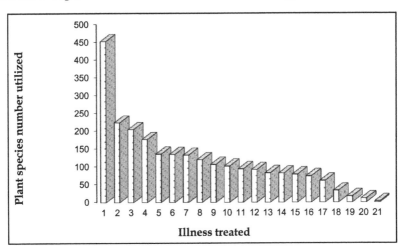

Fig. 2. Species number used to treat different illness: 1 digestive, 2 trauma, 3 symptomatic, 4 respiratory, 5 female sex, 6 urinary, 7 muscle-skeletal, 8 nerve, 9 circulatory, 10 metabolic, 11 nonspecific, 12 "cultural" cold or fright, 13 children, 14 poisoning, 15 infectious, 16 treatment, 17 malignancies, 18 eye, 19 ear, 20 male sex, 21 psychiatric.

It should be noted that acute respiratory illnesses were the leading cause of morbidity in people of Morelos between 2000 and 2009. During the same period intestinal infections occupied the 2nd place of morbidity (SSM 2011). People interviewed on the use of plants have traditional knowledge that allows them to use a more accessible and less expensive to address two of the leading causes of illness. Therefore, it is important to spread knowledge and use of such species. The identity of the species and medicinal plants represent the indicators that explain the most frequent morbidity and mortality in the population. We reiterate the importance to spread knowledge and use of native species which are also useful to many people. We recorded a few plants for the treatment of diseases of the circulatory system, diabetes or malignancy. This situation is related to the recent evolution of these diseases in the communities.

3.3 Use of firewood

Obtaining energy from wood is still a key activity for family subsistence. In Mexico, 80% of the inhabitants of rural communities generate energy from the firewood (Balvanera et al., 2009). In addition, extraction of firewood is a source of income for people living in areas with a climate that has a markedly seasonal rainfall, as in the dry season cannot cultivate the land. This activity is the only source of income for Cuauchichinola Limón households. This community lives in poverty and the collection of firewood is the only source of income for families, for a period of 5 to 7 months, each year. To evaluate the effect of the removal of wood on the tree community, it was assumed a gradient of extraction, which would increase the intensity of collection of firewood with closeness to the households. We worked with randomized blocks in a pseudo-replication. The distance for each block is delimited depending on the frequency distribution of the extraction trips. We counted in blocks and measured for diameter stumps and the same procedure was applied to the tree community to have an area index in both cases. Extraction variables were the number of stumps, and the number of standing trees plus the number of stumps. The indicator of disturbance caused by the extraction equals the number and area of the stumps in relation to the total recorded in the tree community.

The collectors selected 18 of the 47 tree species recorded for use as firewood, based on criteria such as hardness when cutting and burning characteristics. Disturbance indices show a gradient in the density of the tree community concentrating higher density in the far side of the community. Five species density changed on this gradient. Firewood extraction implies a possible conservation value supported by the lore, for biological reasons: using the species and plant stage development with the best quality for combustion, plants of seed dispersal anemocorous and without dormancy, to ensure prompt germination, development and recovery of the trees felled. Another good response of traditional knowledge is timely and not causes disturbance or riot-fell catastrophic, which minimally impacts, similar to a falling tree in the forest, which retains the seral stage of vegetation, and finally, with leading order of collection which begins with the branches before proceeding to the whole tree logs, and branches includes half of the tree biomass. In this way, the community is provided with firewood but the tree has opportunity to re-growth and recovering its structure. The benefits referred to as selective logging, promote density (Bruenig, 1991) and dominance of certain tree species. In conclusion, the appropriation of firewood is an activity that reveals the lore of the inhabitants of "El Limón" on the tree community around them. This knowledge translates into conservation management practices that could be incorporated into a plan for sustainable forestry, based on the active participation of the inhabitants of this community.

4. Studies at local-regional scale

4.1 Method

Indicators of current status of traditional ecological knowledge were estimated indirectly, to investigate how people maximize the utilization of biodiversity and the number of options for attaining survival, from plant resources availability. For the development of the indicators we elaborated a database of plants used traditionally consulting bibliographic information about six case studies that describe the traditional use of plants from the viewpoint of resource inventory. We selected these dimensions in order to homogenize and compare the six selected cases since there are some differences in the kind of traditional ecological knowledge registered by each author. Although sometimes was possible search complementary dates to homogenize information using the floristic list of plant used, such is the case of the origin of the plants, we thought that not always is correct. For example, some authors give information about the distribution of the use of the different plant species during a year; another describes the status of conservation of the vegetation where the plants are collected or classify the management degree of the plants used as cultivated, tolerated, or promoted. In these examples, we have some of the specific characteristics of local management which corresponds to a particular cultural, environmental and technological development framework and should not be inferred from a different local framework. The ethnobotanical information was classified based on the next five dimensions designed to describe the maximization of the diversity and number of options for the use of plant resources (see Equation 1-13):

Dimension 1. Use of families, genera and species of useful plants

Indicators estimated based on the territorial extension.

These indicators will provide a measure of the availability of useful resources per unit area, which allows comparisons with other studies and at different geographical scales (local, municipal, state, national). It is also an indirect measure of the potential of traditional ecological knowledge to generate goods and services and its condition. A higher value would indicate a greater availability of useful resources.

$$\frac{\text{Number of botanical families with wild and cultivated useful species}}{\text{Municipal area}} \qquad (1)$$

$$\frac{\text{Number of genera with wild and cultivated useful species}}{\text{Municipal area}} \qquad (2)$$

$$\frac{\text{Number of wild and cultivated useful species}}{\text{Municipal area}} \qquad (3)$$

$$\frac{\text{Number of plant families with wild useful species}}{\text{Municipal area}} \qquad (4)$$

$$\frac{\text{Number of botanical genera with wild useful species}}{\text{Municipal area}} \tag{5}$$

$$\frac{\text{Number of botanical wild useful species}}{\text{Municipal area}} \tag{6}$$

Indicators estimated based on the botanical richness

These indicators give us a measure of the richness of families, genera and plant species, which is being used locally. This scale is used for comparison because it is more likely to use the resources in the locality. A greater tendency to use diversity could be a proxy for the state of conservation of traditional ecological knowledge, as when resources are best known to have the potential for exploitation. Consider that among the limitations of these studies, nor floristic inventories, nor can ethnobotanical be fully known.

$$\frac{\text{Number of botanical families with wild and cultivated useful species}}{\text{Number of inhabitants in the community}} \tag{7}$$

$$\frac{\text{Number of botanical genera with wild and cultivated useful species}}{\text{Number of inhabitants in the community}} \tag{8}$$

$$\frac{\text{Number of wild and cultivated useful species}}{\text{Number of inhabitants in the community}} \tag{9}$$

$$\text{Ratio of plant families used} =$$
$$\frac{\text{Number of botanical families with wild useful species per location}}{\text{Number of families with wild plants registered in the Morelos state}} \times 100 \tag{10}$$

$$\text{Ratio of botanical genera used} =$$
$$\frac{\text{Number of botanical genera with wild useful species per location}}{\text{Number of genera with wild plants registered in the Morelos state}} \times 100 \tag{11}$$

$$\text{Ratio of plants used} =$$
$$\frac{\text{Number of wild useful species per location}}{\text{Number of wild plants registered in the Morelos state}} \times 100 \tag{12}$$

Dimension 2. Use as many native species

Proportion of native useful species

Proportion of useful introduced species

The use of a greater proportion of useful species would show that the introduced species replace the use of local flora. The value from the relationship between native and wild shows how many wild plants are used for each one introduced species. Therefore, if we obtain greater numbers, the relevance of native plants is greater. The replacement of native species by introduced species is interpreted as a proxy for the replacement of local traditional ecological knowledge and the reduction of plant resources available. If this were the case, the generation of options to satisfy the necessities for survival of the community will be diminishes.

$$\frac{\text{Number of useful native species}}{\text{Number of useful introduced species}}$$

Dimension 3. Maximizing the number of uses assigned to the species

Proportion of useful species by number of uses.

This indicator is an indirect measure of the conservation status about traditional ecological knowledge. It is expected that the number of useful plants was greater where there is greater preservation of knowledge. In areas where there is less conservation of traditional ecological knowledge we would expect the absence of species with numerous uses. Therefore, this is an indicator of the state of conservation of traditional ecological knowledge. The same indicator shows the potential of traditional ecological knowledge to solve more than a basic need.

Dimension 4. Integral utilization of species

Proportion of useful species by the number of plant structures used.

Integral use will be achieved if we know the morphology of the plant, and the uses of each part. In this sense, the presence of plants from which it takes more than one part indirectly indicates a certain level of conservation of traditional ecological knowledge.

Dimension 5. Maximizing use of available species in the wild

Proportion of botanical families with useful species wild and cultivated

Ratio of botanical genera with useful species wild and cultivated

Proportion of wild and cultivated species useful

These indicators show a greater reliance on indirect natural environment for goods and services. A greater number of goods and services would be a good indicator of the state of conservation of traditional ecological knowledge. Obtaining goods and services should be persistent throughout the year. The trend towards temporary lack of goods and services is likely to involve loss of traditional knowledge.

Some of the dimensions included are related with the diversity of plants distributed in the communities selected (use of families, genera and species of useful plants; use as many native species; maximizing use of available species in the wild). Actions should be taken to control the direct and indirect factors that affect the preservation of diversity and its sustainable use (Millennium Ecosystem Assessment, 2005; Challenger & Dirzo, 2009). For example, the social-political factors should be implemented in order to avoid the replacement of forest with urban colonization and/or the economic incentives that promote the establishment of monocultures. The highest value of these dimensions shows indirectly that authorities of a community, region or country are been conducted towards sustainable development. Biophysical, social, and economic variables can influence on the indicators (Figure 3). Thus, the six case studies were selected based on the differences of the

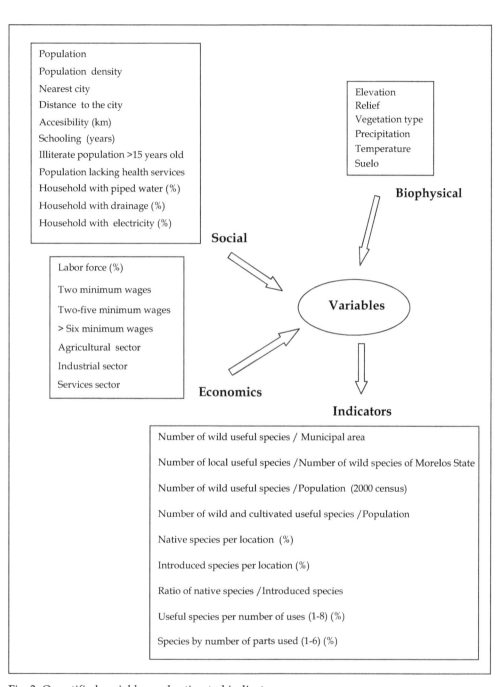

Fig. 3. Quantified variables and estimated indicators.

communities in the physical and biological attributes related to human activity such as accessibility, proximity to urban centers or different socioeconomic characteristics. The socioeconomic characterization of the population was done based on data recorded in the censuses of population and housing from which we obtained information on economically active population occupied in agriculture, livestock and forestry, as well as industrial and services; income, illiterate population, lacking health services, availability of electricity, piped water and drainage in the household. The official census do not include information on the contribution of traditional ecological knowledge to meet essential needs or medicinal, food, implicit in the welfare of low income families.

The relation between the indicators and the physical and socioeconomic variables was established using multivariate statistics. The indicators useful to use multivariate statistics, pre-selected from a correlation analysis between variables to avoid co-linearity, are listed below:

Indicators to maximize the diversity and number of options.

Number of families, genera and species per unit area.

Number of families, genera and species used in relation to the number of families, genera and species of flora recorded in the state.

Indicators of utilization of as many native species as possible.

Number of native species useful in relation to the number of useful introduced species.

Native wild species used

Indicators to maximize the number of uses assigned to species.

Number of uses per species.

Distribution of the number of species per number of uses.

Indicators of the total utilization of the species

Number of plant structures used.

Distribution of the number of species in relation to the number of plant parts used.

Indicators to maximize use of available species in the wild.

Proportion of botanical families with useful species wild and cultivated

Ratio of botanical genera with useful wild and cultivated species

Proportion of wild and cultivated useful species

Indicators to maximize the number of uses assigned to species.

Number of uses per species.

Distribution of the number of species per number of uses.

Indicators of total utilize the species

Number of plant parts used.

Distribution of the number of species in relation to the number of number of plant parts used.

Indicators to maximize use of available species in the wild.

Number of species that produce goods and services per month of the year.

Goods and services type produced by species and month of the year.

4.2 Results. Plant biodiversity maximization and the number of options

We will address the study related to the maximization of plant biodiversity and the number of options to ensure minimum subsistence risk. The information corresponds to six different locations of the state of Morelos, Mexico, with different biophysical attributes, in relation to human activity such as accessibility, proximity to urban centers, or other socio-economical traits (Figure 3). We analyzed them using multivariate methods (ordination and clustering), following Rohlf (2000).

The principal component analysis (Figure 4) explained 90% of the total variation for the first three components. The variables contributing the most in principal component 1 were among the socio-economic: the level of schooling of the population, people lacking access to health care, population density, and the mean wage of 2-5 per inhabitant. The most relevant

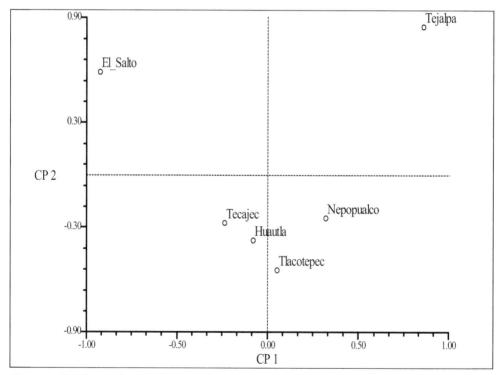

Fig. 4. Ordination of the rural communities based on plant utilization and socio-economical attributes.

variables pertaining to plant use were: species of plants with two types of use, number of plant parts used, ratio between the number of useful wild species and the number of inhabitants, and the ratio between the number of useful wild and cultivated species and the number of inhabitants. The communities are arranged in a gradient defined by the level of development, where El Salto and Tejalpa correspond to the least and the highest socio-economic development. The latter has been an urban area near by the state capital, having the largest population, access to health systems and higher rates of education. In contrast, El Salto has the highest number of useful wild species in relation to population, and where the wild plants have a greater number of uses. The importance of using native species in the El Salto and Tejalpa relates primarily to the existence of Natural Protected Areas in their vicinity. El Salto is immersed in the Reserve Biosphere Sierra de Huautla, while Tejalpa borders the State Protected Natural Area "The Texcal." The importance of using native plants in Tejalpa is also related to the origin of its inhabitants, as though there are many immigrants, whose origin for most of them is Tlapa, Guerrero, a community that has the same type of vegetation (tropical deciduous forest) as the one in the Texcal (Monroy & Ayala 2003). It also has pre-Hispanic cultural roots.

For the Principal Component 2, the most relevant variables of the socio-economic group were: households with electricity and distance to the nearest city, while in the group of variables of use highlighted: the relationship between native and introduced species, the percentage of native species per location, and the percentage of introduced species per location. In the second axis, the communities were arranged in a gradient that primarily involves attributes related to the knowledge of plants, since towards the upper end of axis 2 are located communities with the greatest percentage of native species per location, in contrast , they contain the lowest percentage of introduced species. Tejalpa has the largest population, percentage of population with income from 2 to 5 minimum wages and access to health services (53%). In contrast, El Salto had only 5% of the population which earns more than minimum wage.

Figure 5 shows a grouping formed by Tlacotepec, Nepopualco, Tecajec and Huautla, having a higher similarity in both, socioeconomic and the use of wild plants. In the latter case highlights the fact that they have similar values in the number of useful species used, the percentage of native species per location, and the higher rates of introduced species. On the other hand, El Salto and Tejalpa are distantly related to the previous group, which may influence mainly the socio-economic traits.

It should be noted, that the contribution of the indicators to characterize the situation of utilization of resources based on traditional knowledge. Is very clear and reflects the social situation that influences on the use of plants to satisfy their needs. It also shows the inverse relationship between acculturation and knowledge of plant biodiversity, environment or the use of adaptive management of resources.

From the sustainability of the resources, joint analysis of economic factors and indices on the traditional ecological knowledge, we might suggest the relevance of the islands of vegetation, in the wild and urban, for the people of the communities having traditional ecological knowledge on the use of plants. It seems important the access that people should have to these islands of vegetation since they will continue using the existing resources even on protected natural areas.

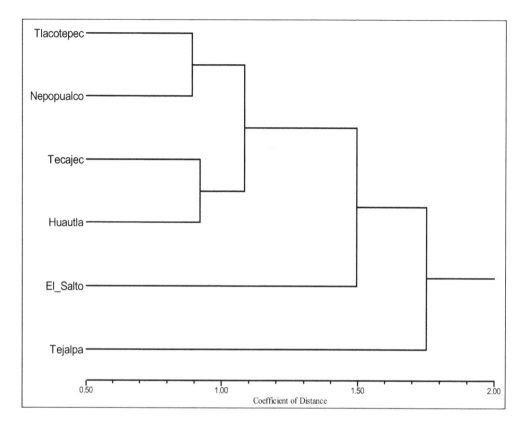

Fig. 5. Grouping of the communities studied in terms of socio-economic and utilization of wild plant species.

5. News pertaining to traditional knowledge

It is necessary for the planning of urban development to consider the need for islands of wild vegetation, either as a source of goods and services or teaching space on the uses and forms of exploitation of resources. Also is required to keep the cover in rural areas. As well as protect and promote traditional knowledge in different areas, e.g., in the education sector should ensure the exchange of knowledge on the resources at the classroom level. In the urban environment field, should be consider the use and practices of resource use of the migrants when planning and design management of existing vegetation islands, as well as other green spaces such as public parks and bed of flowers. Whereas in urban areas is still preserved traditional knowledge about plants, it is necessary to protect and promote this kind of knowledge in education, environment and health sectors, which is necessary, to promote the informed participation of the population in decision-making in these areas. What is new with respect to traditional knowledge is currently being recognized the benefit of traditional ecological knowledge and thought in their legal protection, based on the agreed Nagoya Protocol in Japan October 29, 2010, which has a binding implications and aims at equitable sharing of benefits arising from genetic resources for communities to conserve their resources based on community standards and participate in innovative features (compartiendosaberes.org, 2011).

6. Limitations of the studies based on traditional knowledge

One of the drawbacks in treating ecological data supported by traditional knowledge is empiricism that fall into subjectivity, which is true for ethnic and rural populations. Scientist rely on concepts like adaptive management that validates a coordinated use between the production process and environmental conditions. An integrated view about the three concepts used in this paper: traditional ecological knowledge, indicators, and sustainability, anticipates that the information is still partial, requiring a participatory process, from diagnosis through the indicators, and systematic monitoring of the processes change in both plant communities and in the forms of appropriation of species, in order to maintain the desired productivity. On the research side, another of the limitations of the proposed indicators is the lack of available information for their estimation, or the time required to conduct the study that may lead to determine the useful plants of any community. However, there are inventories of the plants used in develop and under develop countries; in addition, there are methods of inventorying relatively fast.

An additional limitation may arise from the variability in selected information sources, by the method of obtaining and recording of traditional ecological knowledge in each of the case studies. For example, in some studies were selected informants according to their degree of knowledge, sometimes the researcher lived in the community while doing their job, but there are those who planned a series of visits over a period of time, and so on. These variations affect the quality and quantity of ethnobotanical information that could gather and therefore, the estimated indexes and their subsequent analysis.

Likewise, we must consider that due to economic constraints and time, inventories of useful plants are not necessarily complete. In the same vein, we note the lack of local floristic inventories which could make a better comparison of the potential of the resources utilized.

7. Future research

Traditional knowledge necessarily requires scientific validation, since there is a tendency to give importance to the potential to combine both approaches to the same problem, to highlight and protect cultural and biological diversity.

The breadth of topics related to traditional knowledge provides the opportunity to have more indicators; for example, those directly related to the use of such plants as the best time for a medicinal use, the criteria for selecting the best type of wood, the characteristics of wood to use in making a chair, and so on. This variety of indicators could be related to different areas of government policy. Example, traditional knowledge related to the use and enjoyment of food and medicinal plants would be linked to the health sector to propose, promote and regulate the consumption of those nutritious vegetables or effective in the treatment of illness. Also, it is required to investigate the potential of this information provided by the indigenous and rural people to be incorporated into formal education (Figure 6), so that helps students to have better nutrition and health care. Similarly, it is necessary to review the legal framework for considering the inclusion of these foods and medicines, utilized traditionally.

Fig. 6. School education children learning about the importance of traditional ecological knowledge.

Future research must be supported in objectives to reach the desired achievement of certain events, including modernity without losing traditions, out of the discourse and influence the sustainability of resources as a means of maintaining ecosystem health and continuity of productive resources, integrating the communities that really have the traditional ecological knowledge. Ensure that economic activities do not exceed the carrying capacity, or social events such as migration, which contribute to the loss of ecological knowledge of ancient tradition and the cultural erosion. Encouraging multilateral political and social decisions based on cultural diversity. Legally protect traditional knowledge for communities to achieve benefits, as recently discussed by the FONCICYT Consortium (compartiendosaberes.org), which also proposed to strengthen and building capacity to articulate traditional and scientific knowledge in innovation processes and resolution of conflicts associated with the management, ownership and use of resources.

8. Conclusion

The maintenance of productive and healthy natural processes emphasizes the value of sustainability, which, if based on traditional ecological knowledge of those communities, processes and experiences through various measures and indicators of performance , would help to better comprehension of the differences of multiple community expectations pertaining to the management systems, in our case, forestry. The synergy of activities between ethnic groups and rural population, as conservationist of the traditional ecological knowledge and the scientific community as adviser in the current situation of resources and proposing solutions, facilitating a practical impact on the utilization and conservation of resources, and the continuity of productive and healthy ecosystem, ensuring   to sustainability.

In Morelos, it is undeniable that there is a wealth of plant species and traditional ecological knowledge still exists, which supports forms of appropriation of resources that contribute to the sustainable production, conservation and maintenance of ecosystem integrity, providing good and services to society over time. The indicators derived from the traditional ecological knowledge show convergence and complementarities with the scientific knowledge. There is also the potential of biodiversity and traditional ecological knowledge, to consider the active participation of local people in the generation, implementation and monitoring of development proposals that seek sustainability and social welfare. Lacks enhance these convergences of knowledges into actions for sustainable development.

9. References

Balvanera, P., H. Cotler et al. (2009). Estado y tendencias de los servicios ecosistémicos. In: *Capital natural de México, Vol. II: Estado de conservación y tendencias de cambio*, R. Dirzo, R. González & I.J. March, (Eds.), 185-245, Comisión Nacional para el Conocimiento y Uso de la Biodiversidad, ISBN 978-607-7607-08-3, México.

Begossi, A.; Hanazaki, N. & Tamashiro, J. (2002). Medicinal plants in the Atlantic forest (Brazil): knowledge, use, and conservation. *Human Ecology*, Vol.30, pp. 281-299, ISSN 0300-7839.

Bennett, B.C. & Prance, G. (2000). Introduced plants in the indigenous pharmacopoeia of northern South America. *Economic Botany*, Vol.54, pp. 90-102, ISSN 0013-0001

Berkes, F.; Colding, J. & Folke, C. (2000). Rediscovery of traditional ecological knowledge as adaptive management. *Ecological Applications*, Vol.10, No.5, pp. 1251-1262, ISSN 1051-0761

Boege, E. (2009). El reto de la conservación de la biodiversidad en los territorios de los pueblos indígenas. In: *Capital natural de México, Vol. II: Estado de conservación y tendencias de cambio*, R. Dirzo, R. González & I.J. March, (Eds.), 603-649, Comisión Nacional para el Conocimiento y Uso de la Biodiversidad, ISBN 978-607-7607-08-3, México.

Boltvinik, J. & Hernández- Laos, E. (2000). *La pobreza y distribución de ingreso en México. 2a edición*, ISBN 968232193X, México

Bruenig, E.F. (1991). Pattern and structure along gradients in natural forests in Borneo and in Amazonia: their significance for the interpretation of stand dynamics and functioning. In: *Rain forest regeneration and management,* A.Gómez-Pompa, T.C. Whitmore & M. Hadley, (Eds.), 235-240, Man and the biosphere series-UNESCO-The Parthenon Publishing Group., ISBN 92-3-102647-X, Paris, Francia.

Caceres, D. M. (2009). La sostenibilidad de las explotaciones campesinas situadas en una reserva natural de Argentina central. *Agrociencia* Vol.43, pp. 539-550, ISSN 1405-3195

Challenger, A. & Dirzo, R. (2009). Factores de cambio y estado de la biodiversidad, In: *Capital natural de México, Vol. II: Estado de conservación y tendencias de cambio,* 37-73, CONABIO, ISBN 978-607-7607-08-03, México

Compartiendo saberes. (2011). Protocolo de Nagoya, vínculos complejos entre equidad, biodiversidad y derechos de propiedad, Available from http://www.compartiendosaberes.org/

CSD Commission on Sustainable Development. (2000). Indicators of sustainable development: guidelines and methodologies. United Nations. p. 266, Available from http://www.un.org/esa/sustdev/isd.htm

Dale, V. H. & Beyeler, S. C. (2001). Challenges in the development and use of ecological indicators. *Ecological Indicators* Vol.1, pp. 3–10, ISSN 1470-160X

Gispert, M. & Rodríguez, H. (1998). *Los Coras: plantas alimentarias y medicinales de su ambiente natural,* Dirección General de Culturas Populares, Instituto Nacional Indigenista, ISBN 970-18-1614-5, México

Halffter, G. & Moreno, C. E. (2005). Significado biológico de las diversidades alfa, beta y gamma. In: *Sobre Diversidad Biológica: El Significado de las Diversidades Alfa, Beta y Gama,* G. Halffter, J. Soberón, P. Koleff & A. Melic, (Eds.), 5-18, m3m: Monografías Tercer Milenio, ISBN 84-932807-7-1, Zaragoza, España

Heink, U. & Kowarik, I. (2010). What are indicators? On the definition of indicators in ecology and environmental planning. *Ecological Indicators,* Vol.10, pp. 584-593, ISSN 1470-1604

Horwich, R.; Lyon, J. & Bose, A. (2011). What Belize Can Teach Us about Grassroots Conservation. *Solutions,* Vol.2, No.3, pp. 51 -58, Available from http://www.thesolutionsjournal.org

Millennium Ecosystem Assessment. (2005). Ecosystems and Human Well-being: Biodiversity Synthesis. World Resources Institute, Washington, DC., Available from http://www.maweb.org/en/Synthesis.aspx

Monroy, R, & Ayala, I. (2003). Importancia del conocimiento etnobotánico frente al proceso de urbanización. *Etnobiologia,* Vol. 3, pp. 79-92, ISSN 1665-2703

Monroy, R. & Monroy-Ortiz, C. (2003). "Saber popular" alternativa mexicana para conservar el bosque tropical caducifolio, *XII Congreso Forestal Mundial. A. Bosques para la gente.* FAO. Québec, Canada, Available from http://www.fao.org/DOCREP/ARTICLE/WFC/XII/0594-C1.HTM

Monroy-Ortiz, C. & Castillo P. (2007). *Plantas medicinales utilizadas en el estado de Morelos*. Comisión Nacional para el Conocimiento y Uso de la Biodiversidad-Universidad Autónoma del Estado de Morelos, ISBN 968-878-277-7, Cuernavaca, México

Monroy-Ortiz, C. & Monroy, R. (2006). *Las plantas, compañeras de siempre. La experiencia en Morelos*, Comisión Nacional para el Conocimiento y Uso de la Biodiversidad-Universidad Autónoma del Estado de Morelos-Comisión Nacional de Áreas Naturales Protegidas-Centro de Investigaciones Biológicas, ISBN 968-878-242-4, Cuernavaca, México

Monroy-Ortíz, C.; García-Moya, E.; Romero-Manzanares, A.; Sánchez-Quintanar, C.; Luna-Cavazos, M.; Uscanga-Mortera, E.; González-Romero, V. & Flores-Guido, J. S. (2009). Participative generation of local indicators for conservation in Morelos, México. *International Journal of Sustainable Development & World Ecology*, Vol.16, No.6, pp. 381-391, ISSN 1350-4509

Muniz de Medeiros, P.; Santos de Almeida, A.L. & da Silva, T.C. (2011). Pressure Indicators of Wood Resource Use in an Atlantic Forest Area, Northeastern Brazil. *Environmental Management*, Vol.47, pp. 410–424, ISSN 0364-1524

Muthcnick, P. A. & McCarthy, B C. (1997). An ethnobotanical analysis of the tree species common to the subtropical moist forests of the Petèn, Guatemala. *Economic Botany*, Vol.51, pp. 158-183, ISSN 0013-0001

Natcher, D C. & Hickey, C. G. (2002). Putting the community back into community-based resource management: a criteria an indicators approach to sustainability. *Human Organization*, Vol.61, No.4, pp. 350-363, ISSN 0018-7259

ONU-Comisión Mundial sobre el Medio Ambiente y el Desarrollo.(1992). *Nuestro futuro común*, ONU, CMMAD, Madrid, España

Ostrom, E.; Burger, J.; Field, C.F.; Norgaard, R.B. & Policansky, D. (1999). Revisiting the commons: local lessons, global challanges. *Science*, Vol. 284, pp. 278-282, ISSN 0036-8075

Persha, L.; Agrawal, A. & Chhatre, A. (2011). Livelihoods, and biodiversity conservation social and ecological synergy: Local rulemaking, forest. *Science*, Vol. 331, pp. 1606-1608, ISSN 1095-9203, DOI: 10.1126/science.1199343

Reyes-García, V. (2009). Conocimiento ecológico tradicional para la conservación: dinámicas y conflictos. *Papeles*, Vol.107, pp. 39-55

Rohlf, F.J. (2000). *NTSYSpc, Numerical Taxonomy and Multivariate Analysis System Version 2.1, User Guide*, Exeter Software, ISBN 0-925031-30-5

SSM. (2011). Anuarios Estadísticos, Secretaría de Salud. Morelos, México: Servicios de Salud de Morelos, Available from http://www.ssm.gob.mx/anuarios.html (Retrieved February 14, 2011)

Toledo, V. M.; Ortiz-Espejel, B.; Cortés, L.; Moguer, P., & Ordóñez, M. D. J. (2003). The multiple use of tropical forests by indigenous peoples in Mexico: a case of adaptive management. *Conservation Ecology*, Vol.7, No.3,):9, Available from http://www.consecol.org/vol7/iss3/art9.

UN (United Nations). (2009). *Información para la adopción de decisiones. Programa 21, Sección IV. Medios de Ejecución, capítulo 40*. United Nations Department of Economical and Social Affairs, Division for the Sustainable Development, Available from http://www.un.org/esa/dsd/agenda21_spanish/res_agenda21_40.shtml

UNESCO. (2010). *Informe Mundial de la UNESCO*. Invertir en la diversidad cultural y en el diálogo intercultural. París, Francia.

9

Conjunctive Use of Surface Water and Groundwater for Sustainable Water Management

H. Ramesh and A. Mahesha

Dept. of Applied Mechanics & Hydraulics,
National Institute of Technology Karnataka, Surathkal, Mangalore
India

1. Introduction

A critical problem that mankind had to face and cope with is how to manage the intensifying competition for water among the expanding urban centers, agricultural sectors and in-stream water uses. Water planner can achieve a better management through basin-wide strategies that include integrated utilization of surface and groundwater which may be defined as conjunctive use (Todd, 1956). Conjunctive use is the simultaneous use of surface water and groundwater. Investment in conjunctive use raises the overall productivity of irrigation systems, extends the area effectively commanded, helps in preventing water logging and can reduce drainage needs. Lettenmaier and Burges (1982) distinguished conjunctive use which deals with the short term use from the long term discharging and recharging processes known as cycle storage. Until late 1950s, development and management of surface water and groundwater were dealt separately, as if they were unrelated systems. Although the adverse effects have been evident, it is only in recent years that conjunctive use is being considered as an important water management practice.

Conjunctive use of surface and groundwater is not a new concept but it has been in practice since last three decades. The term 'conjunctive' used here is to integrate surface and groundwater resources. It includes interaction between surface water and groundwater through groundwater recharge, hydrological cycle, water balance components etc. These parameters will be used for modeling the groundwater flow and its interaction with surface water. Buras (1963) used dynamic programming to determine design criteria and operating policy for a conjunctively managed system supplying water to agricultural fields. Chun et. al., (1964) used a simulation model to examine alternative plans for conjunctive operation of surface water and groundwater in California, USA. Dracup (1965); Longenbaugh (1970) and Milligan (1969) developed a parametric linear programming model for a conjunctive surface water and groundwater system in southern California, USA.

A GIS linked conjunctive use groundwater – surface water flow model (MODFLOW) was done by Ruud et al, (2001); Sarwar (1999). An overview paper on conjunctive use of surface water and groundwater was presented by Wranchien et al, (2002) giving more emphasis to holistic approach of management. The interaction between surface and groundwater was also studied by various authors elsewhere (Sophocleous, 2002; Ozt et al, 2003; LaBolle et al, (2003). A simple groundwater balance model was developed (Peranginangin et al, 2004) based on 15–20 years (1980–1999) of hydro-meteorological, land use, soil and other relevant data to generate the hydro-geologic information needed for the water-accounting procedure in conjunctive use. A regional conjunctive use model was developed by Rao et al, 2004; Schoups et al, 2005 for a near-real deltaic aquifer system irrigated from a diversion system with some reference to hydro-geoclimatic conditions prevalent in the east coastal deltas of India. A numerical model for conjunctive use surface and groundwater flow was developed and alternating direction implicit method was applied for model solution (Chuenchooklin et al, 2006).

2. Conjunctive use optimization

Optimization techniques were introduced by Castle and Linderborg (1961), who formulated a linear programming to allocate water from two sources (surface water and groundwater) to agricultural areas. Due to the development of advanced digital computer and optimization technique, later, a dynamic programming model (Aron 1969) developed to determine the optimum allocation of surface water and groundwater. Yu and Haimes (1974) discussed hierarchical multi-level approach to conjunctive use of surface water and groundwater systems, emphasizing hierarchical decision making in a general sense. Integrated Groundwater and Surface water Model (IGSM) was first developed by Yan and Smith (1994) at the University of California, USA. The major constraint in IGSM is the semi-explicit time descretization and its incapability that fails to properly couple and simultaneously solve groundwater and surface water models with appropriate mass balance head convergence under practical conditions. An extensive examination of the literature covering conjunctive use of groundwater-surface water summarized chronologically (Maknoon and Burges, 1978); Miles and Rushton (1983); (McKee et al, 2004), reveals in nearly all cases that the analysis of conjunctive use was dominated by one or several parameters which were extensively modeled.

The optimization models were developed by Menenti et al, (1992) and Deshan, (1995) and Karamouz et al, (2004) to allocate optimum water for agricultural benefits in the river basins. Effective use of groundwater simulation codes as management decision tools requires the establishment of their functionality, performance characteristics and applicability to the problems at hand (Paul et al, 1997). This is accomplished through systematic code-testing protocol and code selection strategy. The protocol contains two main elements: functionality analysis and performance evaluation. Functionality analysis is the description and measurement of the capabilities of a simulation code; performance evaluation concerns the appraisal of the code's operational characteristics (e.g., computational accuracy and efficiency, sensitivity for problem design and parameter selection and reproducibility). Testing of groundwater simulation codes may take the form of (1) benchmarking with known independently derived analytical solutions; (2) intra-comparison using different code functions inciting the same system responses; (3) inter-comparison with comparable simulation codes; or (4) comparison with field or laboratory experiments. The results of the various tests are analyzed using standardized statistical and graphical techniques to identify performance strengths and weaknesses of code and testing procedures. The solution of optimization model was done by dynamic programming. A multi-stage decision model was developed by Azaiez (2002) for the conjunctive use of groundwater and surface water with an artificial recharge. He assumed certain supply and a random demand and an integrated opportunity cost explicitly for the unsatisfied demand. He also incorporated the importance of weight attributed by the decision-makers to the final groundwater level at the end of the planning horizon. An integrated hydrologic-economic modeling framework for optimizing conjunctive use of surface and groundwater at the river basin scale (Velázquez et al, 2006).

3. Conjunctive use modeling options

Conjunctive use modeling of surface water and ground water has wide applications in water resources management, ecology, eco-hydrology and agricultural water management. Conjunctive use model are developed based on the purpose and objective. Conjunctive use model are developed based on the technique used and may be classified as :

- Simulation and prediction models,
- Dynamic programming models,
- Linear programming models,
- Hierarchical optimization models,
- Nonlinear programming models and others.

Simulation approaches provide a framework for conceptualizing, analyzing and evaluating stream–aquifer systems. Since the governing partial differential equations for complex heterogeneous ground water and stream–aquifer systems are not amenable to closed form analytical solution, various numerical models using finite difference or finite element methods have been used for solution) simulation and optimization models and decision-support tools that have proven to be valuable in the planning and management of regional water supplies (Chun et al., 1964; Bredehoeft and Young, 1983, Latif and James, 1991; Chaves-Morales et al., 1992; Marino, 2001).

The system dynamics, initially developed by Jay W. Forrester (Forrester 1961), uses a perspective based on information feedback and mutual or recursive causality to understand the dynamics of complex physical, biological, social, and other systems. In system dynamics, the relation between structure and behavior is based on the concept of stock-flow diagrams. The process of model development, combining program flowchart with spatial system configuration, provokes modeler can build model easily. System dynamics is a computer-aided approach to evaluate the interrelationships of components and activities within complex systems. The most important feature of this approach is to elucidate the endogenous structure of the system under study, to see how the different elements of the system actually relate to one another, and to experiment with changing relations within the system when different decisions are included. Dynamic programming (DP) has been used because of its advantages in modeling sequential decision making processes, and applicability to nonlinear systems, ability to incorporate stochasticity of hydrologic processes and obtain global optimality even for complex policies (Buras, 1963; Aron, 1969; Provencher and Burt, 1994). However, the "curse of dimensionality" seems to be the major reason for limited use of DP in conjunctive use studies as it considers physical system as lumped.

Linear Programming (LP) has been the most widely used technique in conjunctive use optimization models.. However, nonlinearities may arise due to the physical representation of the system or the cost structure for surface and groundwater use. For example, Stream-aquifer interaction can be represented by a linear function of stream stage and groundwater elevation where groundwater level is at or above the streambed. However, the stream stage is a nonlinear function of discharge or reservoir release.

Hierarchical optimization was first defined by Bracken and McGill (1974) as a generalization of mathematical programming. In this context the constraint region is implicitly determined by a series of optimization problems which must be solved in a predetermined sequence. Hierarchical optimization models were developed and applied in conjunctive use by Maddock (1972, 1973); Yu and Haimes (1974) and Paudyal and Gupta (1990).

Non linear programming models: The solution of a conjunctive use problem with nonlinear constraints because of very complex and some parameters are non linear. Hence such a model is called nonlinear conjunctive use optimization model. E.g,. In order to solve the conjunctive use problem, the ground water flow and mass transport models will need to be run numerous times that the problem may not be solvable (Taghavi et al. 1994). E.g., groundwater quality problems and groundwater head constraint.

Despite the many different optimization models and techniques that have been applied, most conjunctive use optimization work reported in the literature deal with hypothetical problems, simple cases or steady state problems. The lack of large-scale complex real world conjunctive use optimization studies is probably due to the great size of the problem resulting when many nodes-cells and long time periods are under consideration for modeling groundwater flow and the interaction between surface and groundwater. Most conjunctive use models reported are created "ad hoc" for a particular problem. Water resources engineers and scientists around the world are trying to develop the different kind of conjunctive use models based on purposes and objectives.

Following are some of the conjunctive use models.

- A simple groundwater balance model
- A GIS linked conjunctive use groundwater – surface water flow model (MODFLOW)
- Interaction of surface water and ground water modeling,
- Integrated Groundwater and Surface water Model (IGSM)
- Conjunctive use optimization model
- Linear optimization model
- Non-linear optimization models
- Multi objective conjunctive use models

Apart from the methods of development of conjunctive use models, there is lot of scope in conjunctive use modeling options. Here is some of the conjunctive use modeling options.

- Surface water and groundwater interaction model.
- Managing soil salinity through conjunctive use model
- Groundwater pumping through conjunctive use model.
- Irrigation water management in command area through conjunctive use model.
- Optimal crop planning and conjunctive use of surface water and groundwater.
- Crop scheduling, nutrients and agricultural water management through conjunctive use model.
- Surface water modeling and management
- Groundwater recharge estimation,

- Optimal allocation of surface water and groundwater in a basin.
- Climate change on surface water and groundwater through conjunctive use model., etc.

4. Conceptual conjunctive use model

The conceptual model of the surface water and groundwater was developed at catchment scale (after Sarwar, 1999) and shown in figure 1. The surface water model was developed based on simple water balance which accounts for input and outputs in the system causing change in storage. The water balance is based on law of conservation of mass. The objective of this model was to find the net groundwater recharge in the basin and this net recharge will be the input to the groundwater model. Hence, mathematically one can represent water balance in a basin as

$$I - O = \pm \Delta S_t \tag{1}$$

where I = total inflow, O = total outflow, ΔS_t = change in groundwater storage.

The conjunctive use surface water and groundwater model was developed based on the concept of hydrologic cycle. It consists of three sub-models viz. surface water model, groundwater model and optimization model. An attempt has been made to bring all the three models under one theme. The conceptual model of the present research is presented in figure 2. The surface hydrological processes follow the law of conservation of mass and are modeled using the water balance. This is identified as surface water model. Out of the infiltrated (net recharge) water into the soil, some percentage contributes to the base flow/subsurface flow and rest of it contributes to the aquifer recharge. The quantum of recharge depends mainly on geo-morphological, soil and hydro-geological parameters. The process of flow of water through the porous media is conceptualized as groundwater model.

Due to increased pressure on water resources (domestic, industrial and agricultural), the equilibrium of these two resources gets affected. So the use of surface water in conjunction with the groundwater may play a significant role in marinating the equilibrium and sustainability of the related system. The detailed descriptions of all the three models are given in the subsequent sections.

Over-exploitation of groundwater causes many problems like groundwater table depletion, water quality degradation and sea water intrusion in coastal areas. This is mainly because of shortage of surface water storage resources and the high investment required for storage. The solution for these challenging tasks may be sought through an optimization model. Usually a conjunctive use optimization model has socio-economic and hydraulic constraints. But in the present study, only hydraulic constraints like maximum allowable groundwater level and maximum stream flow utilization were taken into account to satisfy the demand (domestic and agricultural) leading to the optimal utilization of both surface water and groundwater. The three models represented in the conceptual model leads to a Decision Support System (DSS) where a suitable decision would be taken considering optimal utilization of water resources.

The model will help the decision makers, policy makers, practicing engineers and agricultural scientists to prepare the action plans for the overall development in the basin. The plausible policies and action plans should be sustainable water supply schemes for both

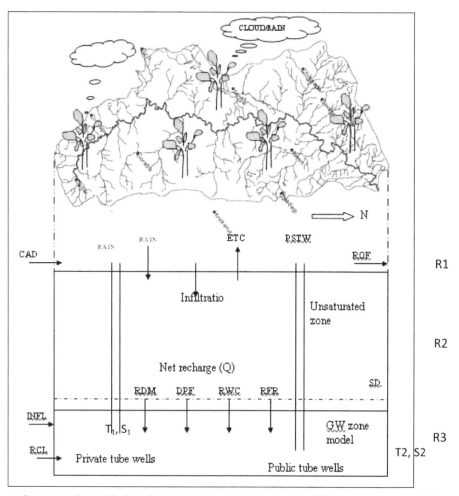

Fig. 1. Conceptual model of surface water & groundwater (modified after Sarwar, 1999).

domestic and agricultural sector in terms of groundwater pumping / surface water utilization to avoid over-exploitation and wastage of energy. It may also involve water resources development plans like construction of recharge structures to compensate for the groundwater level depletion, adoption of suitable cropping pattern and crop schedule to achieve better yield and economy with available water resources in the basin.

The net recharge to the groundwater will be computed by integrating the water balance elements considering R1 and R2 together. The flow in the saturated zone i.e. (groundwater reservoir R3) will be simulated using the groundwater model. The net recharge of a catchment area is then given by

$$Q = RFR + DPF + RDM + RWC + RCL + INFL - ROF - ET_a - EFL$$
$$- PSTW - PPTW - SD \tag{2}$$

where

Q = Net recharge to the aquifer	RFR = Recharge from rainfall
DPF = Deep percolation from field	RDM=Recharge from distributory & minors
RWC = Recharge from water courses	RCL = Recharge from link canals
INFL = Inflow from adjacent area	ROF = Surface runoff
ET_a = Crop evapo-transpiration	EFL = Evaporation from fallow/ bare soil
PSTW = Pumpage by public tube wells	PPTW = Pumpage by private tube wells
CAD = Canal deliveries	SD = Seepage from water table to surface drains

It is assumed that, there is no interflow from adjacent areas into the catchment. Also, the basin is assumed to be geologically and hydrologically single system. The above equation is not applicable every where, suitable modification can be done to suit the interested area by considering all the above components or deleting some of the components.

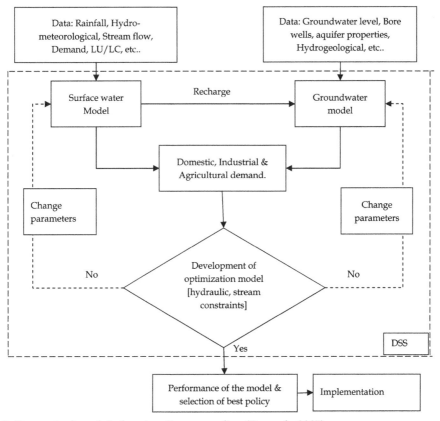

Fig. 2. Conceptual model of conjunctive use policy (Ramesh, 2007)

5. Conjunctive use model

5.1 Computations of water balance components

Water balance components will be compute using available standard models. The modeler can look in to suitable model and the inputs for such models should be supplied through field experiments. Some of models of water balance equation are given below.

5.1.1 Recharge from rainfall (RFR)

Many rainfall recharge models are available to estimate recharge from rainfall. One of the methods is regression based model to estimate recharge from rainfall.

5.1.2 Deep Percolation from field (DPF)

This component is to be estimated based on gross draft plus additional recharge of 5% (GEC, 1997). To estimate groundwater draft, an inventory of wells and a sample survey of groundwater draft from various types of wells (state tube wells, private tube wells and open wells) are required. For state tube wells, information about their number, running hours per day, discharge, and number of days of operation in a season is available in the concerned departments. To compute the draft from private tube wells, pumping sets and rates etc., sample surveys have to be conducted regarding their number, discharge and withdrawals over the season.

5.1.3 Recharge from distributaries (RDM)

It can be estimated separately for lined and unlined canals. Suitable loses can be used or estimated values form past studies can be used. As reported by the Indian Standards (IS 9452, 1980), the loss of water by seepage from unlined canals in India varies from 0.3 to 7.0 m^3/sec / million square meter of wetted area. It is calculated by the following relation:

$$\text{Losses in } m^3 / \sec / km = \left\{\frac{C}{200}\right\} * (B+D)^{\frac{2}{3}} \tag{3}$$

where B=bed width, D=depth of water in meters, C=constant varies from 1 for intermittent to 0.75 for continuous.

As per GEC (1997) recommendations:

i. for unlined canals in normal soils
    -   1.8 to 2.5 m^3/sec / million square meters of wetted area
ii. unlined canals in sandy soils with some silt content
    -   3 to 3.5 m^3/sec / million square meters of wetted area.

5.1.4 Recharge from water courses (RWC)

Recommendations made by GEC, India (1997) are based on average water spread area. Recharge from storage tanks and ponds may be taken as 1.4 mm per day for the period in which tank has water. If the data on average water spread area is not available, then 60% of the maximum water spread area may be used. Recharge due to check dams and nala bunds may be taken as 50% of gross storage.

5.1.5 Surface runoff (ROF)

Direct runoff in a catchment depends on soil type, land cover and rainfall. Of the many methods available for estimating the runoff from rainfall, the curve number method (USDA-SCS, 1964) is the most popular. The curve number method makes use of soil categorization based on infiltration rates and land use i.e., the manner in which the soil surface is covered and its hydrologic conditions are important parameters influencing the runoff. The advantage of this method compared to other methods lies in the fact that the parameters used here are relatively easy to estimate. The final empirical equation given by USDA-SCS (1964) is as follows:

$$Q = \frac{(P-I_a)^2}{[S+(P-I_a)]} \tag{4}$$

where   Q - actual runoff, P – rainfall, I_a- initial abstraction and S - Potential maximum retention after runoff begins which is expressed in terms of Curve Number (CN) given by the relation.

$$S = \frac{25400}{CN} - 254 \tag{5}$$

The parameter CN depends on a combinations of hydrologic soil, vegetation and land use complex (SVL) and antecedent moisture condition of a watershed. But this method has been modified by the Ministry of Agriculture, India (1972) to suite Indian conditions. The initial abstraction (I_a) is usually taken as equal to 0.2S for Indian conditions. Hence, equation (4) becomes

$$Q = \frac{(P-0.2S)^2}{P-0.8S} \tag{6}$$

The Curve Numbers for different SVL and AMC condition can be taken from Handbook of Hydrology (Ministry of Agriculture, India, 1972).

5.1.6 Crop Evapo-transpiration (ET$_a$)

This is the major loss in the water balance studies. It is the combined loss of water in the form of evaporation from soil surface / water and the transpiration from plant or vegetation. It can be calculated by the following equation as suggested by FAO (1956)

$$ET_a = K_C * ET_0 \tag{7}$$

Where ET_a = evapo-transpiration of specific crop (L/T)
ET_0 = potential / reference crop evapo-transpiration (L/T)
K_c= crop coefficient (dimensionless)

The reference crop evapo-transpiration is estimated according to Penman-Monteith (1980) equation.

$$ET_0 = \frac{0.408\Delta(Rn-G)+\gamma\dfrac{900}{Ta+273}Uz(e_a-e_d)}{\Delta+\gamma(1+0.34U_2)} \tag{8}$$

Where ET_0 = reference evapo-transpiration [mm day^{-1}],
       Rn = net radiation at crop surface [MJ m^{-2} day^{-1}],
       Rnl = net outgoing long wave radiation [MJ m^{-2} day^{-1}],
       Ra = net incoming shortwave radiation [MJ m^{-2} day^{-1}],
       Ra = extra terrestrial radiation [s m^{-1}],
       G = soil heat flux [MJ m^{-2} day^{-1}],
       Ta = average air temperature in deg C,
       U_2 = wind speed at 2 meter height [m s^{-1}],
       e_a = saturation vapour pressure [kPa],
       e_d = actual vapor pressure [kPa], e_s - e_a saturation vapour pressure deficit [kPa],
       Δ = slope of the saturation vapor pressure [kPa °C^{-1}],
       γ = psychrometric constant [kPa °C^{-1}].

5.1.6 Evaporation from fallow and barren soil (EFL)

It is estimated by making use of the following equation:

$$EFL = EPF * FSE(1 - XR) * CCA$$

(9)

where EFL = evaporation from fallow land (L^3/T),
       EPF = equivalent evaporation factor,
       FSE = free surface evaporation (pan evaporation), XR = the ratio of cropped to
       cultivable area, CCA = cultivable command area (L^2), EPF is calculated as

$$EPF = \left[\left(\frac{0.55}{(0.66 + WTD)} \right) + 0.009 \right]$$

(10)

where WTD = depth to water table below soil surface.

5.1.7 Pumpage from tube wells (PTW)

Groundwater Pumpage from private and public tube wells is calculated by the following relation to account for the groundwater abstraction.

$$PTW = 0.083 * NPTW * UTF * AD * TOH$$

(11)

where NPTW = no. of private tube wells,
       UTF = the utilization factor for each month,
       AD = the actual discharge of private tube wells (m^3/sec),
       TOH = total operational hours in a year (hrs), 0.083 = conversion factor

5.2 Development of numerical groundwater model

5.2.1 Model selection

Understanding the physics of groundwater flow and its interaction with surface water is a complex task. This is mainly because of the heterogeneity of the geo-hydrological formation, the complexity in the recharge and the boundary conditions of the aquifer system. Thus the role of numerical models has got utmost importance in the field of aquifer simulation. There are many numerical models available to simulate groundwater system. The numerical models are mainly based on finite difference (FD), finite element (FE), finite volume (FV) and finite boundary (FB) approaches. For many groundwater problems, the finite element method is superior to classical finite difference models (Willis and Yeh, 1987). Heterogeneities and irregular boundary conditions can be handled easily by the finite element method. This is in contrast to difference approximations that require complicated interpolation schemes to approximate the complex boundary conditions. Moreover, the size of element can be easily modified to reflect rapidly changing state variables or parameter values in the finite element method.

5.2.2 Governing equations

The groundwater flow modeling methodology given by American Society for Testing Materials (ASTM) presented in figure 3 was used in the present study.

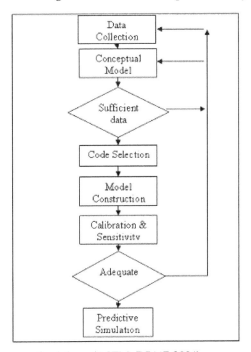

Fig. 3. Groundwater flow methodology (ASTM, D5447-2004)

The groundwater flow in an aquifer is represented by the following differential equations (Jacob, 1963),

For steady state condition:

$$\frac{\partial}{\partial x}\left(T_x \frac{\partial h}{\partial x}\right) + \frac{\partial}{\partial y}\left(T_y \frac{\partial h}{\partial y}\right) \pm G(x,y) = 0 \tag{12}$$

For transient condition:

$$\frac{\partial}{\partial x}\left(T_x \frac{\partial h}{\partial x}\right) + \frac{\partial}{\partial y}\left(T_y \frac{\partial h}{\partial y}\right) = S\frac{\partial h}{\partial t} \pm G(x,y,t) \tag{13}$$

where T_x and T_y are the x and y – direction transmissivities respectively (m²/day);
h- Piezometric head (m); S – Storage coefficient (dimensionless);
G(x,y,t) – Pumping/Recharge (m³/day); t – Time (days);
x & y – Coordinate axes.

5.2.3 Initial and boundary conditions

The initial groundwater level is provided as initial condition.

$$h(x_i,y_i,0) = h\ (x_i,y_i) \tag{14}$$

where, $h(x_i, y_i)$ is initial piezometric head.

The boundary condition is the combination of Dirichlet and Neumann conditions

$$h = \overline{h}(x,y,t)\ \ \text{on } \Gamma_1 \tag{15}$$

and

$$T_x \frac{\partial h}{\partial x}l_x + T_y \frac{\partial h}{\partial y}l_y = q(x,y,t)\ \ \text{on } \Gamma_2 \tag{16}$$

where $\overline{h}$ – specified piezometric head and l_x, l_y, l_z are the direction cosines between the normal to the boundary surface and the coordinate axes; Γ_1 represents those parts of the boundary where h is known and is therefore specified. q is prescribed for the remaining part of the boundary (Γ_2), which is the flow rate per unit area across of the boundary. For the general case of transient flow with phreatic surface moving with a velocity V_n normal to its instantaneous configuration, the quantity of flow entering its unit area is given by

$$q = V_n S + I * l_x \tag{17}$$

where S is the specific yield coefficient relating the total volume of material to the quantity of fluid which can be drained. I is the infiltration or evaporation.

The pumping or recharging well at a particular point in the domain is represented as:

$$Q_h^w(x_i,t) = \sum_m Q_m^w \prod \left\{ \delta\left(x_i - x_i^m\right)\right\} \quad \text{for} \quad \forall(x_i - x_i^m) \in \Omega \tag{18}$$

where Q_h^w = a well function, Q_m^w = pumping or recharge rate of a single well (m³/sec)

X_i^m = coordinate of well (m)

5.2.4 Finite element formulation

The finite element solution of equations (12 & 13) with initial and boundary conditions (14 - 16) is derived using Gelarkin's weighted residuals method. The Galerkin finite element method is a widely used technique for sub-surface flow simulations due to its efficiency and suitability (Pinder and Grey, 1977). The variable h is approximated as

$$h \approx \hat{h} = \sum_{i=1}^{n} N_i h_i \tag{19}$$

Over the domain; where N_i are the interpolation functions; h_i are the nodal values of h; n is the number of nodes.

The application of Galerkin method to the steady state equation yields following integral equation:

$$\int_\Omega RN_i d\Omega = 0 ; \qquad i = 1, 2, \dots n \tag{20}$$

in which

$$R = \left[\frac{\partial}{\partial x}\left(T_x \frac{\partial h}{\partial x} \right) + \frac{\partial}{\partial y}\left(T_y \frac{\partial h}{\partial y} \right) \right] \pm G(x,y) \tag{21}$$

where Ω refers to the area of flow domain

By applying Green's theorem, equation (20) can be modified to

$$\int_\Omega \left(\frac{\partial N_i}{\partial x} \sum_1^n T_x \frac{\partial N_j}{\partial x} + \frac{\partial N_i}{\partial y} \sum_1^n T_y \frac{\partial N_j}{\partial y} \right) h_i d\Omega -$$
$$- \int_\Gamma N_i \left(\sum_1^n T_x \frac{\partial h}{\partial x} l_x + \sum_1^n T_y \frac{\partial h}{\partial y} l_y \right) d\Gamma \pm \int_\Omega N_i G_i d\Omega = 0 \tag{22}$$

where Γ refers to external boundary. Equation (22) leads to a system of simultaneous equations which can be expressed as

$$[P]\{h\} = \{F\} \tag{23}$$

where [P] – conductivity matrix; {h}- vector of nodal values; {F} – load vector

$$P_{ij} = \sum \int_E \left(\frac{\partial N_i}{\partial x} T_x \frac{\partial N_j}{\partial x} + \frac{\partial N_i}{\partial y} T_y \frac{\partial N_j}{\partial y} \right) d\Omega \tag{24}$$

and

$$F_i = \sum \int_{\Gamma E} N_i q d\Gamma \pm \int_E N_i G_i d\Omega \tag{25}$$

where E denotes an element; ΓE refers to elements with an external boundary. The element equations are assembled into global system of equation. The prescribed boundary conditions are inserted at this stage and the solution is obtained using Gauss elimination routine.

5.2.5 Development of transient model

Rewriting the equation (13) describing linearized unsteady groundwater flow

$$T\left\{\frac{\partial^2 h}{\partial x^2}+\frac{\partial^2 h}{\partial y^2}\right\}-S\frac{\partial h}{\partial t}=\pm G(x,y,t) \tag{26}$$

To solve this equation, homogeneous and isotropic domain with boundary Γ in the time interval $(0, t_n)$ is assumed. Both an essential and natural boundary conditions are imposed on the boundary.

$$h(x,y,t)=h_0(x,y,t) \ \text{on} \ \Gamma_1 \tag{27}$$

$$T\left(\frac{\partial h}{\partial x}l_x+\frac{\partial h}{\partial y}l_y\right)+q_0=0 \ \text{on} \ \Gamma_2 \tag{28}$$

where l_x and l_y are directional cosines of the outward normal to Γ. h_0- specified piezometric head; q_0- specified flux

The following initial condition is imposed on the domain Ω

$$h(x,y,0)=H(x,y) \ \text{in} \ \Omega \tag{29}$$

where H – Initial piezometric head. Applying the Galerkin method to equation (26),

$$\int_\Omega N_i\left\{T\left[\frac{\partial^2 h}{\partial x^2}+\frac{\partial^2 h}{\partial y^2}\right]-S\frac{\partial h}{\partial t}\pm G(x,y,t)\right\}d\Omega=0 \qquad i=1,2,3....n \tag{30}$$

where n is number of nodes in the finite element mesh and N_i are the shape functions. Now applying Green's theorem yields

$$\int_\Omega T\left\{\frac{\partial N_i}{\partial x}\frac{\partial h}{\partial x}+\frac{\partial N_i}{\partial y}\frac{\partial h}{\partial y}\right\}d\Omega-\int_{\Gamma_2}N_iq_0d\Gamma+\int_\Omega N_iS\frac{\partial h}{\partial t}d\Omega\pm\int_\Omega N_iG_id\Omega=0 \tag{31}$$

The resulting system can be conveniently written in matrix form:

$$[P]\{h\}+[L]\left\{\frac{\partial h}{\partial t}\right\}=-\{F\} \tag{32}$$

where [P] - conductivity matrix; [L] - storativity matrix.

The elements of the matrices are given as

$$p_{ij}=\int_{\Omega^e}T\left\{\frac{\partial N_i}{\partial x}\frac{\partial N_j}{\partial x}+\frac{\partial N_i}{\partial y}\frac{\partial N_j}{\partial y}\right\}d\Omega \tag{33}$$

$$l_{ij} = \int_{\Omega^e} SN_iN_j d\Omega \qquad (34)$$

$$f_i = \int N_i q d\Gamma \pm \int N_i G dA \qquad (35)$$

For linear triangular element shown in figure 4, the interpolation function is given as

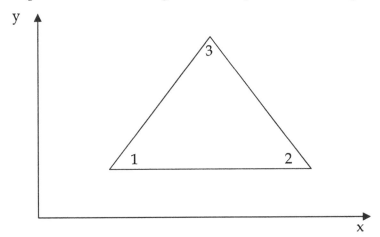

Fig. 4. Linear triangular finite element

$$N_i = \frac{(a_i + b_i x + c_i y)}{2A} \qquad (36)$$

the element matrix is given below

$$\left[K^e\right]\{h^e\} =$$

$$= T \int_{\Omega^e} \left(\begin{Bmatrix} \dfrac{\partial N_1}{\partial x} \\ \dfrac{\partial N_2}{\partial x} \\ \dfrac{\partial N_3}{\partial x} \end{Bmatrix} \begin{Bmatrix} \dfrac{\partial N_1}{\partial x} & \dfrac{\partial N_2}{\partial x} & \dfrac{\partial N_3}{\partial x} \end{Bmatrix} + \begin{Bmatrix} \dfrac{\partial N_1}{\partial y} \\ \dfrac{\partial N_2}{\partial y} \\ \dfrac{\partial N_3}{\partial y} \end{Bmatrix} \begin{Bmatrix} \dfrac{\partial N_1}{\partial y} & \dfrac{\partial N_2}{\partial y} & \dfrac{\partial N_3}{\partial y} \end{Bmatrix} \right) d\Omega \{h^e\} \qquad (37)$$

where Ω^e is the element domain

performing integration after substituting the shape functions we get

$$\left[K^e\right] = T \begin{bmatrix} k_{11} & k_{12} & k_{13} \\ k_{21} & k_{22} & k_{23} \\ k_{31} & k_{32} & k_{33} \end{bmatrix} \qquad (38)$$

in which

$$k_{11} = \frac{1}{4A}\left[(x_3 - x_2)^2 + (y_2 - y_3)^2\right] \tag{39}$$

$$k_{12} = \frac{1}{4A}\left[(x_3 - x_2)(x_1 - x_3) + (y_2 - y_3)(y_3 - y_1)\right] \tag{40}$$

$$k_{13} = \frac{1}{4A}\left[(x_3 - x_2)(x_2 - x_1) + (y_2 - y_3)(y_1 - y_2)\right] \tag{41}$$

$$k_{21} = k_{12} \tag{42}$$

$$k_{22} = \frac{1}{4A}\left[(x_1 - x_3)^2 + (y_3 - y_1)^2\right] \tag{43}$$

$$k_{23} = \frac{1}{4A}\left[(x_1 - x_3)(x_2 - x_1) + (y_3 - y_1)(y_1 - y_2)\right] \tag{44}$$

$$k_{31} = k_{13} \tag{45}$$

$$k_{32} = k_{23} \tag{46}$$

$$k_{33} = \frac{1}{4A}\left[(x_2 - x_1)^2 + (y_1 - y_2)^2\right] \tag{47}$$

where A is the area of triangle.

Considering forward difference scheme for the time derivative term in the equation (32)

$$\frac{\partial h}{\partial t} = \frac{h_{t+1} - h_t}{\Delta t} \tag{48}$$

Also, considering the system of equations marching with time, a time stepping scheme is introduced with a factor θ. The solution accuracy and the numerical stability depends the choice of values of θ, is of decisive significance. Most frequently 1, ½, or 0 are substituted for θ. Equation (32) thus obtains the form:

Case (i): θ =1, Backward scheme;

$$(L + P_{t+1}\Delta t)h_{t+1} = Lh_t - F_{t+1}\Delta t \tag{49}$$

Case (ii): θ =1/2, central (Crank-Nicolson) scheme

$$\left(L + \frac{1}{2}P_{t+1}\Delta t\right)h_{t+1} = \left(L - \frac{1}{2}P_t\Delta t\right)h_t - \frac{1}{2}(F_t + F_{t+1})\Delta t \tag{50}$$

Case (iii): $\theta = 0$, forward scheme;

$$Lh_{t+1} = (L - P_t\Delta t)h_t - F_t\Delta t \qquad (51)$$

In the present study, an implicit scheme with $\theta = 1/2$, (Crank-Nicholson scheme) was adopted. The model was operated on a monthly basis to suit the availability of data. A computer code was developed in Visual C++ for the entire process and programme is given in appendix II. The results are presented in GIS platform (ESRI, 2004) for better visualization.

5.2.6 Model calibration

In the present study, trial and error calibration (figure 5) procedure is adopted. Initially, the aquifer parameters such as transmissivity (T) and storativity (S) are assigned based on the field test results. The simulated and measured values of piezometric heads were compared by adjusting the model parameters to improve the fit. The recharge components were varied within the range presented in table.

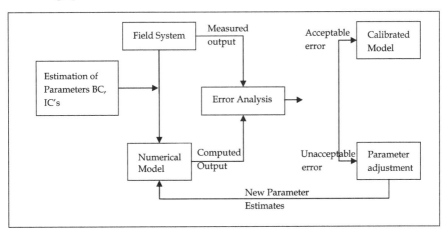

Fig. 5. Trial and error calibration procedures (Anderson and Woessner, 1992)

The following input parameters have received particular attention during the calibration.

- Specific yield/storage coefficient, transmissivity of aquifer.
- Factor for recharge from distributaries and minors
- Factor for recharge from watercourse
- Factor for discharge to surface drains
- Factor for recharge from rainfall
- Factor for evaporation from watercourse surfaces and bank vegetation

For the second and third parameters, an empirical equation has been used to compute recharge to groundwater depending on the depth of rainfall as discussed earlier. For the rest of parameters, following generalities of transient calibration (Boonstra and Ridder, 1990) were followed.

- First, change the input parameters for those areas where the largest deviation occurs

- Change one type of input parameter in each run
- Determine whether any change of input parameter in one area will have positive or negative effect in other areas.

5.2.7 Convergence criteria

A modeler must decide what levels of accuracy are appropriate for comparative assessment of alternatives. A generalized model with limited accuracy doesn't provide the required level of confidence in the selection of a water management strategy, while beyond certain limits that is required to provide a rational basis for comparing alternatives is wasteful. This can be achieved by imposing convergence criterion and tolerance limits in the model to stop the number of iterations. The following convergence criterion is used in the present study.

$$\frac{\sqrt{\sum_{j=n}^{n} h_i^2} - \sqrt{\sum_{j=1}^{n} h_{i-1}^2}}{\sqrt{\sum_{j=1}^{n} h_i^2}} \le \varepsilon \tag{52}$$

where i = iteration index, j = no. of nodes, ε = tolerance limit (0.001).

The attainment of steady sate is also monitored using the above relationship with 'i' representing the time level.

5.2.8 Model validation

The objective of model performance analysis is to quantify how well the model simulates the physical system and to identify the problem if any, in the model. The method typically used to quantify model error is to compute the difference between predicted and observed values of piezometric heads (Residual) at the measuring location. The scatter diagrams, together with computed coefficient of determination indicate where the greatest discrepancies occur and whether there are few major discrepancies or general disagreement between predictions and observations (Karlheinz and Moreno, 1996).

The performance of the calibrated model could be quantified by a number of statistics comparing the observed and simulated hydraulic heads (ASTM, 1993). Following measures of the goodness of fit between measured and simulated water levels (Sarwar, 1999) were calculated in this study.

Mean Error (ME)

$$ME = \frac{1}{N} \sum_{i=1}^{N} (P_i - O_i) \tag{53}$$

Root Mean Square Error (RMSE)

$$RMSE = \left[\sum_{i=1}^{N} \frac{(P_i - O_i)^2}{N} \right]^{\frac{1}{2}} \left[\frac{100}{\overline{O}} \right] \tag{54}$$

where P is simulated value, O is the observed value, $\bar{O}$ is the mean observed value and N is the number of observations.

The error parameters, generally used for evaluating the calibration quality (Frey Berg, 1988; Anderson & Woessner, 1992, Madan et al., 1996) are to be tabulated.

5.3 Development of conjunctive use optimization model

Management optimization is a powerful technique for computing optimal solutions for challenging management problems, such as maximizing quantity of water or minimizing operating costs. The management problem is mathematically formulated to represent the desired objectives of the decision maker (e.g., minimize costs), as well as the associated constraints (e.g., required water supply rate). Algorithms compute the optimal solution (e.g., pumping rates of individual wells) and quantify its sensitivity to various problem components (e.g., cost coefficients, constraint limits, etc.). Surface water and groundwater systems are often intimately connected. Industrial, commercial, and agricultural land uses affect aquifer recharge and discharge, which in turn impact spring discharge to, and seepage from, surface water bodies. Irrigated agriculture is a significant component of river and aquifer water budgets in many areas of the world. Surface water applied in excess of crop consumptive requirements enters the groundwater system increasing aquifer water levels and spring discharge. Groundwater pumping for irrigation or other consumptive uses creates the opposite effect. The Snake river in southern Idaho is a prime example of a surface water system that is greatly affected by groundwater conditions which changes in response to irrigation practices (Miller et al, 2003). Integrated river basin modeling with distributed groundwater simulation and dynamic stream-aquifer interaction allows a more realistic representation of conjunctive use and the associated economic results (Velázquez et al, 2006).

In the present study, optimization problem was formulated as a linear programming problem with the objective of maximizing water production from wells and from streams given by John et al, (2003) with a little modification. The objective function has the following constraints:

1. Maintaining groundwater level at or above specified level.
2. Utilization of stream flow at or below maximum specified rates.
3. Limiting the maximum increase in groundwater withdrawals.

5.3.1 Water demand

The total water demand in the basin is considered to be of domestic, agricultural and industrial sectors. The water demand will be projected over next two decades based on past decadal census data.

Domestic Water Demand

Domestic water demand is the total quantity of water that is being used for drinking, cooking, washing, cleaning etc. therefore it is mainly depending on the number of population. The domestic water demand will be calculated by population forecast based on arithmetic progression, geometric progression, incremental increase and national average.

However, specific assessment of growth potential shall be taken into consideration while arriving at the final population forecast.

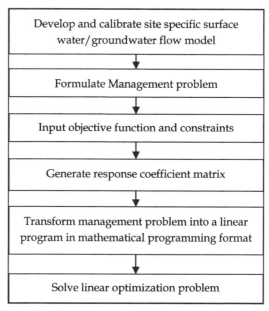

Fig. 6. Flow chart of optimization modeling process (Czarnecki, 2003)

Arithmetic progression: This method is based on the assumption that population increases at constant rate. A constant increment growth is added periodically based on the past records. This method generally gives a low rate of population growth and can be used where growths are not conspicuous.

Population forecast for $P_n = P_{2001} + (X) * n$ (55)

where X= Average population increase / decade
n = No. of decades

Geometrical progression method: In this method, percentage increase or percentage growth rate per decade is assumed to be constant, and the percentage increase is compounded over existing population every decade.   This method normally predicts greater values of population and is used for the areas with scope for huge expansion plans.

Population forecast for $P_n = P_{2001} + (1 + M / 100)^n$ (56)

where M = Average percentage increase in population

Incremental increase method: In this method, the average incremental increase is calculated from the available data. To the present population, the average incremental increase per decade is added and the population of next decade is obtained. Like this, the process is repeated till the population in the desired decade is reached.

Population forecast for $\qquad$ $P_n = P_{2001} + (X + Z) * n$ $\qquad$ (57)

where X = Average population increase / decade = Total increase/ No. of decades
Y = Net incremental increase; Z = Average incremental increase; n = No. of decades.

Final decision on the estimation of domestic water demand shall be based on the realistic projection for the project period using the methods described above. However, the decade growth rate must be limited to 20 percent (National average growth, India), when the projected population growth is more than 20 percent per decade. However, under exceptional circumstances where the growth rate beyond 20 percent, it should be substantiated by data.

5.3.2 Objective function

The objective of the present optimization model is to maximize the water production from both groundwater and the surface water resources. The objective function has the following form:

$$Maximize \cdots Z = \sum q_{well} + \sum q_{river} \qquad (58)$$

where Z – is the total managed water withdrawal in Mm³/day
Σq_{well} – is the sum of groundwater withdrawal rates in Mm³/day
Σq_{river} – is the sum of surface water withdrawal rates from all managed river reaches in Mm³/d.

The following constraints are formulated to solve the objective function.

i.   **Hydraulic head constraints**

This is the constraint imposed based on the groundwater level fluctuation in an aquifer. For achieving sustainability, a critical groundwater level is to be worked out by analyzing. The following hydraulic constraint is to be satisfied for sustainable groundwater management.

$$h_c \leq h_{maximum} \qquad (59)$$

where h_c is the hydraulic head (water level) at the given location c, in meter.

$h_{maximum}$ is the groundwater level altitude at half the thickness of the aquifer in meter. There is a flexibility of fixing h_c in the model based on hydrogeology and groundwater levels fluctuation. The above equation allows an aquifer to drain up to critical hydraulic head (h_c).

ii.   **Stream flow constraints**

stream flow constraint as the maximum utilization of the stream flow within the basin. The stream flow constraint was derived based on simple mass balance equation as follows:

$$q_{head} + \sum q_{overland} \pm \sum q_{groundwater} - \sum q_{diversion} - \sum q_{river} \leq q_{maximum} \qquad (60)$$

where q_{head} is the flow rate into the head of stream in m³/d
$\Sigma q_{overland}$ is the sum of all overland and tributary flow into stream reach in m³/d
$\Sigma q_{groundwater}$ is the net sum of all groundwater flow to or from stream reach R, in m³/d

$\Sigma q_{diversion}$ is the sum of all surface water diversions from stream reach in m^3/d

Σq_{river} is the sum of all potential withdrawal excluding diversions from stream in m^3/d

$q_{maximum}$ is the minimum permissible surface water flow rate for stream in m^3/d

But the data on river head and groundwater are not available in the study area. The selection of maximum stream flow rate ($q_{maximum}$) depends on the downstream requirement. Therefore the above constraint reduces to the following form:

$$\sum q_{overland} - \sum q_{diverion} - \sum q_{river} \leq q_{\max imum} \tag{61}$$

iii. Groundwater pumping limits

If no limits are imposed on the potential amount of water that can be pumped at each managed well, then those wells nearest to the sources of water, such as rivers or general head boundaries will be the first to be supplied water, thus capturing flow that would otherwise reach wells farther from the sources.

$$0 \leq \sum q_{well} \leq mq_{well(year)} \tag{62}$$

where, $\sum q_{wells}$ is the optimal groundwater withdrawal, Mm^3/d

m is a multiplier to account for annual increase in pumping rate

$q_{well (year)}$ is the total amount withdrawn in the particular year from the wells in Mm^3/d.

iv. Surface water withdrawal limits

No limits are imposed on optimized withdrawal from river such that the range in optimal withdrawal was between zero and maximum amount of water available at a given point in a river. This specification permitted the analysis where water could be withdrawal and the maximum quantity available. Withdrawals will be allowed only at one point where river constraint is specified i.e. at measuring point.

5.4 Case Study

A humid, tropical river basin is chosen for the application of conjunctive use model. The Varada river basin of southern India lies between latitude 14° to 15° 15′ N and longitude 74° 45′ to 75° 45′ E (Fig. 7). The river originates at an altitude of 610 m above the mean sea level (MSL) in the western ghats (mountainous forest range parallel to west coast) and drains an area of about 5020 Km². The river flows towards north-east for about 220 Km and joins the river Tungabhadra. Physiographically, Varada basin consists of western ghats on the west and a plateau region in the east. Sirsi, Siddapur, Soraba, Sagar, and part of Hanagal taluks are covered by the western ghat region and form a dense tropical forest zone. The remaining area falls under the plateau region. The average annual rainfall in the western ghat and the plateau regions are 2070mm and 775mm respectively. The rainfall is mainly confined to June to November and the rest of the year is usually a dry season.

5.4.1 Model calibration

Steady state calibration

The basin is discretized into 329 linear triangular elements with 196 nodes (fig.8). The aquifer condition of January 1993 was used as initial condition for the steady state model calibration.   A number of trail runs were made by varying both transmissivity and storativity values of the aquifers so that root mean square (RMS) error was kept below 0.5m. The simulated (computed) versus observed heads for selected observation points (wells), are shown in figure 9.   The figure indicates a good agreement between the simulated and observed water levels. This was also found to true for other observation wells.

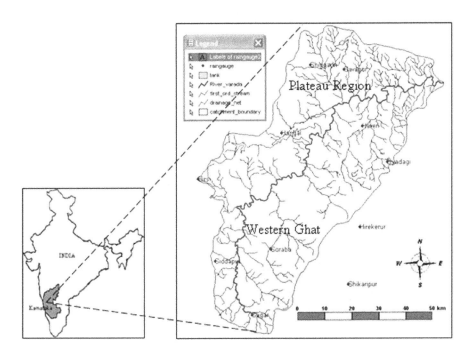

Fig. 7. Study area- Varada Catchment (Ramesh and Mahesha, 2008)

Error Measures	Well Location (nodes)						
	6	14	43	105	139	175	185
ME	-0.08	-0.31	0.47	0.55	-0.21	-0.43	-0.08
RMSE	0.65	0.46	0.73	0.78	0.56	0.76	0.69
R^2	0.83	0.89	0.89	0.89	0.91	0.78	0.86

Table 1. Goodness of fit statistics for comparison between observed and simulated heads (Ramesh and Mahesha, 2008).

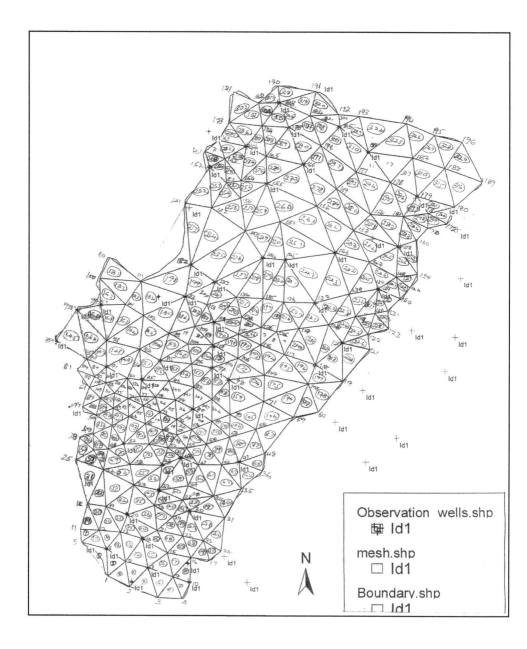

Fig. 8. Finite element mesh for Varada river basin (mod. after Ramesh & Mahesha, 2008)

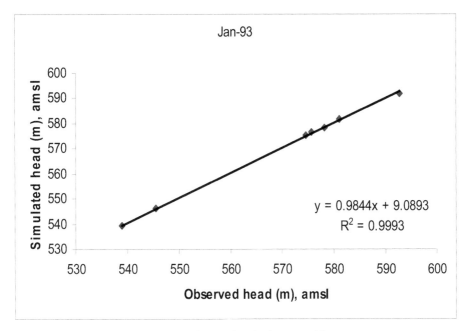

Fig. 9. Observed and simulated groundwater levels during calibration

Transient calibration

The transient calibration was carried out for the period January 1993 to December 1998. The hydraulic conductivity values, boundary conditions and the water levels, arrived through the steady state model calibration were then used as the initial condition in the transient model calibration. These were used along with the storage coefficient distribution and time variable recharge and pumping distribution. A numbers of trial runs were made by varying the storage coefficient (S) values within the observed range so that a reasonably good match was obtained between computed and observed water levels. The transmissivity values are already arrived at during the steady state calibration. Forty seven observation wells were selected as the fitting wells after consideration of their data availability and distribution in the region. The calibrated storage coefficient values for the western ghat zone and plain area zone were found to be 0.0025 and 0.0063 respectively.

The computed well hydrographs for these boreholes show a fairly good agreement with the field values. The disagreement observed in some observation wells (OW-3) is generally attributed to differences in the initial head conditions arrived through steady state calibration, variation in pumping pattern and insufficient bore well data. Nearly 10% of the total bore wells are unauthorized and hence pumping rate and pattern differs from the official data.

5.5 Simulation of predicted scenarios (Groundwater levels)

The calibrated model is applied to predict the basin response over the short term i.e. 2004 – 2010 under various aquifer stress scenarios. The total water demand in the basin was

Year	RFR	RWC	RLC	INFL	ROF+BF	ETc	PTW	Net Recharge (R)	Recharge /Rainfall	% increase PTW	Classification
1993	1200.59	61.28	30.64	2.45	367.54	596.70	98.49	232.23	0.19	0.00	---
1994	1443.55	75.73	37.86	3.03	364.22	596.70	123.51	475.74	0.33	25.40	Wet year
1995	968.67	48.47	24.24	1.94	263.91	596.70	134.61	48.10	0.05	36.67	---
1996	1030.93	53.79	26.89	2.15	210.25	596.70	161.74	145.07	0.14	64.22	---
1997	1487.62	74.68	37.34	2.99	242.84	596.70	183.60	579.49	0.39	86.41	---
1998	1293.58	64.73	32.37	2.59	186.67	596.70	198.87	411.03	0.32	101.92	---
1999	1270.48	66.60	33.30	2.66	241.17	596.70	188.25	346.92	0.27	91.14	---
2000	1232.28	62.04	31.02	2.48	219.11	596.70	181.08	330.92	0.27	83.86	---
2001	713.05	36.32	18.16	1.45	80.61	596.70	164.54	-72.87	-0.10	67.06	Dry Year
2002	936.65	46.83	23.42	1.87	96.71	596.70	203.13	112.23	0.12	106.24	---
2003	902.14	45.41	22.70	1.82	85.76	596.70	207.05	82.55	0.09	110.22	---
Average	1134.50						167.72	244.26	0.19	70.29	Average

Table 2. Water balance components for the period 1993-2003 (in mm)

Year		Jan	Mar	Feb	Apr	May	Jun	Jul	Aug	Sep	Oct	Nov	Dec
1993	Q	0.000	0.546	0.000	0.682	1.692	5.985	9.821	6.867	1.358	9.771	1.463	1.887
	P	1.615	2.936	2.479	3.302	3.734	5.798	7.266	5.078	2.683	9.626	3.467	2.112
1994	Q	0.000	0.081	0.000	2.248	1.081	10.247	19.046	6.491	3.333	6.619	0.302	0.000
	P	1.990	1.719	1.734	1.749	2.698	8.023	15.430	4.992	2.628	6.431	0.776	0.980
1995	Q	0.066	0.000	0.000	1.181	1.015	2.319	12.049	5.335	4.340	4.160	1.210	0.000
	P	2.035	2.035	1.779	2.065	2.902	2.050	12.170	4.358	2.326	5.996	0.962	1.133
1996	Q	0.000	0.000	0.022	1.386	1.109	7.439	9.811	5.669	3.773	4.982	0.297	0.859
	P	2.020	2.065	2.045	2.115	2.574	5.770	12.383	4.279	2.325	5.975	1.031	1.233
1997	Q	0.083	1.768	0.000	1.170	1.039	9.822	13.038	12.507	0.765	4.708	2.689	1.215
	P	2.389	2.136	2.494	2.414	2.079	7.511	11.686	10.470	2.263	5.980	1.165	1.318
1998	Q	0.000	0.004	0.000	0.067	1.733	9.177	10.723	6.821	7.565	4.907	1.413	0.094
	P	2.269	2.404	2.429	2.014	2.113	7.166	10.786	7.809	5.319	5.760	1.182	1.207
1999	Q	0.000	0.018	0.024	0.615	3.657	8.144	17.128	4.089	2.173	7.585	0.054	0.000
	P	0.140	0.601	0.632	0.803	3.165	7.551	17.151	4.375	2.251	7.197	0.380	0.179
2000	Q	0.295	0.000	0.021	0.788	1.283	7.145	11.565	8.680	6.455	4.315	0.069	0.035
	P	0.123	0.573	0.602	0.813	0.961	6.664	10.967	8.755	5.565	3.489	0.410	0.212
2001	Q	3.420	0.000	0.004	1.676	0.718	5.541	5.382	5.264	3.253	1.645	0.229	0.152
	P	0.116	0.582	0.636	1.303	0.778	5.385	5.301	5.029	2.782	1.337	0.395	0.186
2002	Q	0.000	0.651	0.020	1.467	1.891	7.288	4.570	8.234	1.295	5.236	0.054	0.028
	P	0.137	0.711	0.649	1.187	0.952	6.611	4.779	8.097	1.179	4.975	0.405	0.199

Table 3. Recharge (Q) & extraction (P) rates in study area (in Mm³/d)

Node No.	Village	Jan-02	Scenario-2 5 % increase in pumping				Scenario-3 10% increase in pumping			
			Jan-07	Jan-08	Jan-09	Jan-10	Jan-07	Jan-08	Jan-09	Jan-10
2	Yadagigalemane	608.61	606.80	605.08	603.44	601.88	604.99	601.72	598.53	596.66
5	Talaguppa	578.73	578.64	578.54	578.44	578.33	578.55	578.34	576.55	571.25
6	Alahalli	572.79	572.52	572.23	571.94	571.62	572.25	571.65	569.69	566.04
10	Ullur	628.20	625.13	622.22	619.46	616.83	622.07	616.56	612.05	609.25
14	Keladi	577.70	577.58	577.47	577.36	577.25	577.46	577.24	576.32	574.49
16	Bommatti	609.13	606.85	604.69	602.63	600.66	604.58	600.46	598.26	596.53
32	Hosabale	592.09	591.68	591.29	590.92	590.57	591.27	590.54	589.49	586.25
43	Yalsi	574.99	574.72	574.45	574.18	573.91	574.44	573.91	571.91	565.72
49	Tyagali	546.10	542.66	539.05	535.26	531.28	539.23	531.66	499.33	436.29
80	Kuppagadde	565.75	564.84	563.98	563.16	562.36	563.94	562.29	562.72	559.22
92	Isloor	626.63	626.70	626.76	626.84	626.91	626.76	626.90	624.86	613.35
95	Jade	553.35	552.33	551.25	550.12	548.94	551.30	549.05	537.38	511.32
97	Anavatti	542.34	542.26	542.17	542.08	541.98	542.17	541.99	538.42	525.95
105	Agasanahalli	540.05	539.92	539.78	539.64	539.48	539.79	539.50	538.00	529.98
114	Makaravalli	549.35	548.97	548.56	548.14	547.69	548.58	547.73	541.81	525.14
139	Motebennur	576.45	575.90	575.38	574.89	574.43	575.35	574.37	574.09	572.63
144	Adur	545.56	545.19	544.81	544.41	543.99	544.83	544.03	538.72	523.45
179	Haleritti	517.91	517.72	517.51	517.30	517.07	517.52	517.09	512.97	502.62
180	Negalur	513.60	513.47	513.34	513.21	513.07	513.35	513.08	513.03	512.30
185	Yalavigi	596.41	596.36	596.32	596.27	596.22	596.32	596.22	594.51	588.74
189	Outlet	501.08	500.79	500.49	500.17	499.84	500.50	499.87	495.02	481.41

Table 4. Predicted groundwater levels (in m with respect to mean sea level) for the scenarios –2 & 3 (January)

Node No.	Village	May-02	Scenario-2 5 % increase in pumping				Scenario-3 10% increase in pumping			
			May-07	May-08	May-09	May-10	May-07	May-08	May-09	May-10
2	Yadagigalemane	607.54	606.41	605.32	604.28	603.28	605.27	603.20	601.29	599.52
5	Talaguppa	578.86	577.96	577.01	576.02	574.97	577.06	575.07	572.89	570.48
6	Alahalli	570.34	569.82	569.28	568.70	568.10	569.30	568.16	566.90	565.52
10	Ullur	626.53	625.26	624.06	622.93	621.86	624.00	621.74	619.72	617.93
14	Keladi	576.55	576.28	575.99	575.69	575.37	576.01	575.40	574.73	574.00
16	Bommatti	569.25	568.59	567.91	567.20	566.46	567.94	566.53	564.98	563.29
32	Hosabale	589.05	588.77	588.50	588.24	587.99	588.48	587.97	587.50	587.08
43	Yalsi	573.48	573.38	573.27	573.19	573.10	573.28	573.10	572.92	572.76
49	Tyagali	604.61	601.82	598.88	595.80	592.56	599.02	592.87	586.10	578.65
80	Kuppagadde	565.23	564.21	563.23	562.31	561.42	563.18	561.34	559.66	558.14
92	Isloor	626.05	625.90	625.74	625.57	625.40	625.75	625.41	625.04	624.63
95	Jade	552.35	551.12	549.83	548.48	547.06	549.90	547.20	544.23	540.96
97	Anavatti	542.25	542.21	542.17	542.12	542.07	542.17	542.08	541.97	541.85
105	Agasanahalli	540.31	539.97	539.60	539.22	538.82	539.62	538.86	538.02	537.10
114	Makaravalli	547.56	546.76	545.91	545.03	544.10	545.95	544.19	542.24	540.10
139	Motebennur	569.74	569.74	569.74	569.74	569.74	569.74	569.74	569.73	569.73
144	Adur	557.96	557.89	557.82	557.75	557.67	557.83	557.68	557.51	557.32
179	Haleritti	522.46	521.87	521.26	520.61	519.93	521.29	520.00	518.58	517.02
180	Negalur	517.20	516.43	515.62	514.77	513.88	515.66	513.96	512.10	510.04
185	Yalavigi	590.10	589.98	589.85	589.72	589.57	589.86	589.59	589.29	588.96
189	Out let	509.01	508.18	507.32	506.41	505.46	507.36	505.55	503.57	501.38

Table 5. Predicted groundwater levels (in m with respect to mean sea level) for the scenarios – 2 & 3 (May)

Node No.	Village	Sep-02	Scenario-2 5% increase in pumping				Scenario-3 10% increase in pumping			
			Sep-07	Sep-08	Sep-09	Sep-10	Sep-07	Sep-08	Sep-09	Sep-10
2	Yadagigalemane	611.12	610.26	609.43	608.63	607.87	609.39	607.80	606.35	605.00
5	Talaguppa	585.43	584.26	583.03	581.74	580.38	583.09	580.51	577.68	574.56
6	Alahalli	578.13	577.52	576.89	576.21	575.51	576.92	575.58	574.11	572.49
10	Ullur	629.67	628.24	626.89	625.60	624.35	626.84	624.26	621.86	619.62
14	Keladi	578.67	578.43	578.08	577.71	577.32	578.09	577.36	576.55	575.66
16	Bommatti	580.15	579.96	579.44	578.90	578.34	579.46	578.39	577.22	575.94
32	Hosabale	594.20	596.49	596.21	595.94	595.68	596.20	595.66	595.15	594.66
43	Yalsi	577.32	577.30	576.93	576.55	576.15	576.95	576.18	575.34	574.42
49	Tyagali	644.03	640.89	637.59	634.12	630.49	637.74	630.83	623.23	614.87
80	Kuppagadde	565.36	567.45	567.15	566.86	566.57	567.15	566.56	565.99	565.44
92	Isloor	633.22	632.65	632.06	631.43	630.78	632.09	630.84	629.48	627.97
95	Jade	570.13	571.16	570.31	569.43	568.50	570.35	568.59	566.65	564.52
97	Anavatti	543.50	545.05	544.55	544.02	543.46	544.57	543.51	542.34	541.06
105	Agasanahalli	543.72	544.39	544.07	543.74	543.39	544.09	543.43	542.70	541.90
114	Makaravalli	549.83	550.13	549.31	548.45	547.55	549.35	547.64	545.76	543.69
139	Motebennur	570.77	571.59	571.49	571.39	571.29	571.48	571.28	571.10	570.94
144	Adur	561.02	561.42	561.37	561.32	561.26	561.37	561.27	561.15	561.03
179	Haleritti	522.39	522.07	521.72	521.35	520.97	521.74	521.01	520.20	519.32
180	Negalur	514.65	514.18	513.68	513.16	512.62	513.70	512.67	511.53	510.27
185	Yalavigi	598.44	598.38	598.32	598.26	598.20	598.32	598.20	598.90	599.00
189	Outlet	510.80	510.56	510.31	510.04	509.76	510.32	509.79	509.21	508.57

Table 6. Predicted groundwater levels (in m with respect to mean sea level) of the scenarios-2 & 3 (September)

predicted based on the historical data. Various levels of increase in water demand are considered to have different options of water management. The aquifer response under different stress scenarios was studied in order to evolve optimal groundwater extraction along with surface water utilization for the sustainable development of water resources. In all, six different scenarios were considered to evolve the optimal management schemes.

The rainfall and pumping data were analyzed for the last 11 years (table 2). It indicates that there is deficient rainfall of about 6% per year with respect to the normal rainfall (1200 mm) and pumping increases by about 7% every year. The years 1994 and 1997 may be considered as wet years with surplus rainfall of more than 10%. The rainfall deficiency of 40% was observed during 2001 which may be considered as dry year. The recharge and extraction of groundwater was estimated for the last 11 years and the results are shown in table 3. Based on these statistics, six scenarios are predicted for the estimation as follows:

1. 2 % increase in the pumping rate of 2003 every year up to 2010.
2. 5 % increase in the pumping rate of 2003 every year up to 2010.
3. 10 % increase in the pumping rate of 2003 every year up to 2010.
4. 5 % increase in pumping with 2 % increase in recharge rate of 2003 every year up to 2010.
5. 20 % increase in the pumping rate of 2003 every year up to 2010.
6. Increase in recharge rate due to proposed inter-linking of Bedti-Varada river.

The simulated groundwater levels for some of the above scenarios are given in table 4 to 6.

5.6 Optimization model

The ultimate objective of the optimization model is to provide estimates of sustainable yield from both groundwater and surface water. Sustainable yield is defined here as a withdrawal rate from the aquifer or from a stream that can be maintained indefinitely without causing violation of either hydraulic-head or stream flow constraints. The optimization problem was solved by graphical method shown in figure 10 for the year 2003. The optimum withdrawals of surface water and groundwater limits were given in table 8. The amount of surface water and groundwater withdrawals in the feasible region (points 1-5) are indicated here. Table 8 clearly indicates that the total sustainable yield of 11.8 Mm^3/d is possible with conjunctive use of surface water (1.6 Mm^3/d) and groundwater (10.2 Mm^3/d) in the Varada basin. The sustainable yield from groundwater is a function of the withdrawal limit specified which accounts for annual average increase of 5-20% of extraction rates from 2003. The distribution of optimal withdrawal rates with upper limits being specified as 5%, 10 % and 20 percent multiples of 2003 groundwater withdrawal rates is continued for a short period up to 2010. The results are listed in table 9.

Considering the minimum possible growth rate of 5%, the sustainable yield of groundwater and surface water are 4.22 Mm^3/d and 0.422 Mm^3/d respectively (figure 11). Specifying an upper withdrawal limit of 10 percent of the 2003 withdrawal rate and continuing every year (scenario 2), the sustainable yield of groundwater from the basin is 5.81 Mm^3/d (table 9), which was about 3.3 Mm^3/d in 2003. If the upper withdrawal limit is increased to 20 percent annually, the sustainable yield of groundwater from the basin is about 12.88 Mm^3/d (table 9). But this rate violates the hydraulic constraint of h_c = 50m and most part of the basin would be subjected to groundwater mining. Hence this projected increase is not feasible with increase in pumping rate of 20% every year. The only option available is to increase surface water withdrawal i.e. stream flow to cater this growth rate. Now, the upper withdrawal limit of river flow increased to 10 percent of the 2003 withdrawal rate (scenario 4) with 10% increase in groundwater withdrawals. The sustainable yield from groundwater for the basin is 5.81 Mm^3/d (table 9) and the surface water withdrawal is about 0.585 Mm^3/d during the year 2010. However, there is further scope in increasing the river flow withdrawal up to 20% per year. If that would be the case, the sustainable yield from the surface water and groundwater is about 1.289 Mm^3/d and 5.81 Mm^3/d respectively.

The optimal conjunctive use surface water groundwater thus leads to sustainable development of the region within the given constraints. Policy decisions need to be centered around these results while planning the overall water resources development of the region. In this study, the numerical model gave an useful insight into the developmental scenarios for the conjunctive use of surface water and groundwater resources in the Varada river basin.

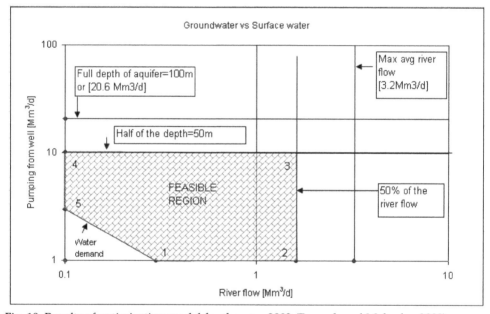

Fig. 10. Results of optimization model for the year 2003 (Ramesh and Mahesha, 2009)

Points	q well [Mm³/day]	q river [Mm³/day]	Z=Σ qwell+Σ qriver [Mm³/day]	Remarks
1	1	0.3	1.3	
2	1	1.6	2.6	
3	10.2	1.6	11.8	Optimum
4	10.2	0.1	10.3	
5	3	0.1	3.1	

Table 8. Optimum withdrawal rates of surface water and groundwater for the year 2003 (Ramesh and Mahesha, 2009)

Sources	Increase in extraction / year	2003	'04	'05	'06	'07	'08	'09	'10	Upper limits
Ground water (wells)	5% increase from 2003 every year	3.00	3.15	3.30	3.48	3.65	3.83	4.02	4.22	10.20
	10% increase from 2003 every year	3.00	3.30	3.63	3.99	4.38	4.81	5.29	5.81	
	20% increase from 2003 every year	3.00	3.60	4.32	5.18	6.22	7.46	8.95	10.74	
Surface water (river)	5% increase from 2003 every year	0.30	0.32	0.33	0.35	0.37	0.38	0.40	0.42	1.60
	10% increase from 2003 every year	0.30	0.33	0.37	0.40	0.44	0.48	0.53	0.59	
	20% increase from 2003 every year	0.30	0.36	0.43	0.52	0.62	0.75	0.90	1.08	

Table 9. Sustainable yield for different upper limits on withdrawals and demand rates [in Mm^3/d]

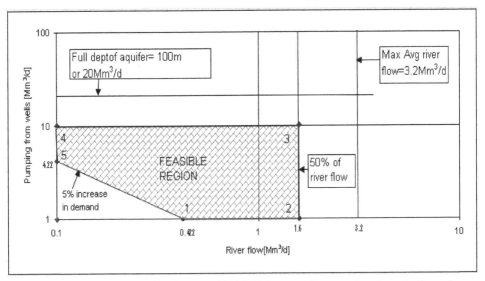

Fig. 11. Results of Scenario-1 for the year 2010 [5% increase in pumping from both surface water & groundwater / year]

5.7 Conclusions

The use of groundwater in conjunction with surface water is gaining prominence in the recent years as a part of water conservation measures in the water stressed regions worldwide. Considering the case study of India, increasing demand for fresh water has put enormous pressure on agriculture and domestic sectors due to population explosion, urbanization, industrialization of expansion of agricultural activities. The agricultural activities in the basin are predominantly controlled by the monsoon rains which are limited to four months in a year and there is an immense need for efficient utilization of available water resources during the rest of the year. To address the increasing demand for fresh water, the government of Karnataka, India is implementing World Bank assisted 'Jal Nirmal' project on the sustainable watershed development programme to ensure supply of safe drinking water to north Karnataka districts. The Varada basin is one of the beneficiaries of the project and is taken up for the present investigation. The results from the study would be useful feedback on the success of the project and the options available for the sustainable development of the region.

An attempt was made in the present study to simulate and allocate the available water resources of the basin for various demands with sustainability approach. The following conclusions may be drawn from the present study:

- The surface water model is based on the water balance approach and the groundwater recharge estimated by it compares well with the other methods. The study evaluated the effect of recharge due to rainfall and other surface water bodies on groundwater through field observations and methods proposed by Groundwater Estimation Committee. The annual average recharge in the basin is estimated to be about 1200 Mm^3
- The numerical solution was effective and accurate enough to simulate the aquifer system with mean error ranging between -0.43 to 0.55 and the correlation coefficient between from 0.78 to 0.91.
- Based on the past records on the increase in freshwater demand, an average increase of 7 % in the groundwater/surface water extraction is estimated. The simulation was carried out to predict the decline in groundwater level for various levels of development. It was predicted that up to 2% increase in extraction rate every year, the system is sustainable. The growth rates more than this may produce undesirable results with groundwater mining.
- The option of surface water supply through run-of-river supply and storage structures may be considered seriously to meet this situation. The present level of river water utilization is 0.2 Mm^3/day which can be increased up to 1.6 Mm^3 /day through adequate canal network and storage structures.
- Operation of additional conjunctive use facilities and storage capacity under flexible water allocation (water transfers) can generate substantial economic benefits to the region. Conjunctive use adds operational flexibility required for water transfers which in tern ensures water allocation flexibility needed to take economical advantage of conjunctive use.
- The optimization model provides a sustainable solution considering different water demands (domestic and agriculture) and available groundwater/surface water resources. Considering a maximum growth rate of 10% every year in the water demand, the optimal conjunctive utilization could be 5.81 Mm^3/day from groundwater resources and 0.585 Mm^3/day from surface water resources. The effective implementation of the developed policies ensures sustainable groundwater development in the study area.
- The proposed Bedti-Varada link system could augment the groundwater/surface water system of the surrounding region significantly even if a minimum utilization of 25% of total transferable amount of 242 Mm^3 is considered.

Scope for Further Investigations

To ensure sustainability, water resources systems need to be planned, designed and managed in such a way as to fully meet the social and economical objectives of both present and future generations and maintaining their ecological, environmental and hydrological integrity. This imposes constraints on every stage of development from project planning to final operation and maintenance. Water Managers and decision makers have to consider a large number of often conflicting demands on the available water and operate water resources systems under numerous social, economic and legal, as well as physical constraints. Economic constraints are equally important in water resources development in a market oriented economy and the concerned agencies may not support it without economic feasibility. In view of this, the present work can be attempted as a nonlinear optimization subjected to the social and economic constraints along with the hydraulic and stream flow constraints. The parameters which will be considered in socio-economic constraints are the gross domestic product (GDP), equity, etc. With the above issues being included, the problem may be viewed as an Integrated Water Resources Management which is the ultimate objective of sustainable development of any region.

6. References

American Society for Testing Materials (ASTM). (2002). *Standard guide for defining boundary conditions in ground-water flow modeling*. ASTM Standard D 5609-94, 4 p.

American Society for Testing Materials (ASTM).2004. Standard Guide for application of a groundwater flow model to a site specific problem. ASTM Standard D 5447-94, 7 p.

Anderson, M. P. and Woessner, W.W. (1992). *Applied groundwater modeling-simulation of flow and advective transport*, San Diego, California, Academic Press, Inc.

Aron, G. (1969). *Optimization of conjunctively managed surface and groundwater resources by dynamic programming*. Univ. of California Water Resources Center. Contrib. No. 129.

Azaiez, M.N., (2002). A model for conjunctive use of ground and surface water with opportunity costs. *European Journal of Operational Research*, 143, 611–624.

Boonstra, J. and Ridder, N.A. (1990). *Numerical modeling of groundwater basins. International institute for Land Reclamation and Improvement (ILRI)*. Publication 29, Wageningen, The Netherlands.

Bracken, J, McGill, J. (1974). Defense applications of mathematical programs with optimization problems in the constraints. *Operations Research* 22: 1086–1096.

Bredehoeft, J.D., Young, R.A. (1983). Conjunctive use of groundwater and surface water for irrigated agriculture: risk aversion. *Water Resour. Res.* 19 (5), 1111–1121.

Buras, N. (1963). Conjunctive operation of dams and aquifers. *J. Hydraulics Div.*, ASCE, 89, HY 6:111.

Castle, E. N. and Linderborg, K. H. (1961). *Economics of groundwater allocation*. Ag. Exp. Sta. Misc. Paper 106, Oregon State University, Corvallis, USA.

Chaves-Morales, J., Marino, M.A., Holzapfel, A.H. (1992). Planning simulation model of irrigation district. *J. Irrig. Drainage Eng., ASCE* 118 (1), 74–87.

Chuenchooklin, S., Ichikawa, T., Patamatamkal, S., Sriboonlue, W. and Kirdpitugra, C. (2006). A conjunctive surface water and groundwater modeling for inundated floodplain in Thailand. *J. Environmental Hydrology*, Vol. 14, pp 1-12.

Chun, R. Y. D., Mitchell, L. R. and Mido, K. W. (1964). Groundwater management for nation's future- optimum conjunctive operation of groundwater basin. J. Hydraulics div., ASCE, 90, HY 4:79.

Czarnecki, J.B., Clark R. B. and Gregory, S.P. (2003). *Conjunctive use optimization model of the Mississippi river valley alluvial aquifer of south eastern Arkansas.* USGS Water Resources Investigation Report 03-4233, USA.

Deshan, T. (1995). Optimal allocation of water resources in large river basins: I Theory. *Water Resources Management*, 9:39-51.

Dracup, J. A. (1965). *The optimum use of a groundwater and surface water system: A parametric linear programming approach.* Univ. California Water Resources Ctr. Contrib. No. 107, Los Angeles., USA.

ESRI, 2004. Environmental System Research Institute, Inc. ArcVieweine desktop-GIS mit den ArcView Extensions.ESRI Gesellschaft for Systemforschung und umweltplanung GmbH, Ringstrabe 7, D-85402.

FAO (1956).Statistical year book, *Food and Agricultural Organization of United Nations*, Rome.

Frey Berg, D.L., 1988. An exercise in groundwater modeling, calibration and prediction. *Ground Water*, 26(3):350-360.

Groundwater Resource Estimation Committee (GEC). 1997. *Groundwater resource estimation methodology*, A Report, Ministry of Water Resources, Government of India, New Delhi.

IS 9452. (1980). *Code of practice for measurement of seepage losses from canals – part 2: inflow outflow method.* Bureau of Indian Standards, New Delhi, India

Jacob, C. E. (1963). *Determining the permeability of water-table aquifers, in Bentall, Ray, compiler, Methods of determining permeability, transmissibility, and drawdown*, U.S. Geol. Survey Water-Supply Paper 1536-1, 245-271.

John B. C., Clark R. B. and Stanton P. G. (2003). *Conjunctive use optimization model of the Mississippi river valley alluvial aquifer of south eastern Arkansas.* USGS Water Resources Investigation Report 03-4233. USA.

Karamouz M., Kerachian R. and Zahraie B. (2004). Monthly water resources and irrigation planning; Case study of conjunctive use of surface water and groundwater resources. *J. Irrigation & Drainage Engg.*, ASCE, 130(5), 391-402.

Karlheiz,S. and Moreno, J. (1996). *A practical guide to groundwater and solute transport modeling.* John Willey and Sons Inc., USA.

LaBolle M. Eric, Ahmed A. Ayman and Fogg E. Graham. (2003). Review of the integrated groundwater and surface water model (IGSM). *Ground Water*, 41(2), 238-246.

Latif, M., James, L.D. (1991). Conjunctive water use to control waterlogging and salinization. *J. Water Resour. Plann. Manage. Div.*, ASCE 117 (6), 611–628.

Lettenmair, D.P. and Burges, S.J. (1982). Cyclic storage: A preliminary assessment, *Ground Water*, 20(3), 278-288.

Longenbaugh, R. R. (1970). Determine optimum operation policies for the conjunctive use of surface water and groundwater using linear programming. *Proc. Special Conference, ASCE – Hydraulic Div.* Minneapolis, USA.

Madan, K. J., Chikamori, K. and Nakarai, Y., (1996). Numerical simulations for artificial recharge of Takaoka groundwater basin, Tosa city, Japan. *Rural and Environmental Engineering J.*, 31(8): 105-124.

Maddock, T. (1973). Management model as a tool for studying the worth of data. *Water Resour. Res.* 9 (2), 270–280.

Maknoon R. and Burges J. S. (1978). Conjunctive use ground and surface water. *Journal American Water Works Association*, 419-424.

Marino, M.A., 2001. Conjunctive Management of Surface Water and Groundwater, Issue 268. IAHS-AISH Publication, pp. 165–173.

McKee, P.W., B.R. Clark and J.B. Czarnecki. (2004). Conjunctive use optimization model and sustainable yield estimation for the Sparta aquifer of southeastern Arkansas and north central Louisiana. *U.S. Geol. Surv., Water Resour. Investigation Rep.*, 03-4231, Little Rock, AR, 30pp.

Menenti, M., J. Chambouleyron, J. Morabito, L. Fornero and L. Stefanini. (1992). Appraisal and optimization of agricultural water use in large irrigation schemes: II Applications. *Water Resources Management*, 6(3), 201-221.

Miles, J. C. and Rushton, K. R. (1983). A coupled surface water and groundwater model. *J. Hydrology*, 62, pp 159-177.

Miller, S. A., Johnson, G. S., Cosgrove, D. M., Larson, R. (2003). Regional scale modeling of surface and ground water interaction in the Snake River basin. *J. American Water Resources Association*, June, 1-12.

Milligan, J. H. (1969). *Optimizing conjunctive use of groundwater and surface water*. Wrt. Res. Lad. Rprt., Utha State University, Logan, USA.

Ministry of Agriculture. (1972). *Hand book of Hydrology*. Soil Conservation Department, Ministry of Agriculture, New Delhi.

Monteith, J.L. (1980). The development and extension of Penman's evaporation formula. In: *Applications of Soil Physics* (D. Hillel, Ed.), Academic Press, New York, 247-253.

Otz H. M., Otz K. H., Otz I. and Siegel I. D. (2003). Surface water / groundwater interactions in the Piora aquifer, Switzerland: Evidence from dye tracing tests. *Hydrogeology Journal*, Vol. 11, 228-239.

Paudyal, G.N., Gupta, A.D. (1990). Irrigation planning by multilevel optimization. *J. Irrig. Drainage Eng.*, ASCE 116 (2), 273–291.

Paul K. M. van der Heijde and Kanzer D. A. (1997). *Groundwater model testing: Systematic evaluation and testing of code functionality and performance*. National Risk Management Research Laboratory, Report No. EPA/600/SR-97/007, United States Environmental Protection Agency, Ada, USA.

Peranginangin N., Ramaswamy S., Norman R. S., Eloise K., Tammo S. S. (2004). Water accounting for conjunctive groundwater/surface water management: case of the Singkarak–Ombilin River basin, Indonesia. *J. Hydrology*, 292, 1–22.

Pinder, J. F. and Gray, W. G. (1977). *Finite element simulations in surface and subsurface hydrology*. Academic Press, New York, 295 pp.

Provencher, B., Burt, O. (1994). Approximating the optimal ground water pumping policy in a multi aquifer stochastic conjunctive use setting. *Water Resour. Res.* 30 (3), 833–843.

Rao, S. V. N., Murty B.S., Thandaveswara, B. S. and Mishra, G. C. (2004). Conjunctive use of surface and groundwater for coastal and deltaic systems. *Journal of Water Resources Planning and Management*, ASCE, 130(3), 255-267.

Ramesh, H. (2007). Development of conjunctive use surface water and groundwater modeling for sustainable development of Varada river basin, Karnataka. *PhD thesis.* Department. of Applied Mechanics & Hydraulics, National Institute of Technology Karnataka, Surathkal, Mangalore 575025, India, 208pp

Ramesh, H and A.Mahesha. (2008). Simulation of Varada aquifer system for sustainable groundwater development, *J. Irrig. & Drain*, ASCE, 134(3), 387-399.

Ramesh, H. and A.Mahesha. (2009). Conjunctive use in India's Varada river basin. *J. American Water Works Assn.*,101:11, 74-83.

Ruud, N. C., T. Harter and A. W. Naugle. (2001). A conjunctive use model for the Tule river groundwater basin in the San Joaquin Valley, California; *Integrated Water Resources Management, Marino, M. A. and S. P. Simonovic* (eds.), IAHS Publication No. 272. 167-173.

Sarwar A. (1999). Development of conjunctive use model, an integrated approach of surface and groundwater modeling using GIS, *PhD Thesis*, University of Bonn, Germany. pp 1-140.

Schoups, G., Addams C. Lee, and Gorelick, S. M. (2005). Multi-objective calibration of a surface water-groundwater flow model in an irrigated agricultural region: Yaqui Valley, Sonora, Mexico. *Hydrol. Earth Sys. Sci.* Discuss., 2, pp 2061–2109.

Sophocleous M. (2002). Interactions between groundwater and surface water: the state of the science. *J.Hydrogeology*, Vol. 10, pp 52-67.

Taghavi, S.A., RE. Howitt and M.A. Marino. (1994). Optimal Control of Ground-Water Quality Management: Nonlinear Programming Approach. *J. of Water Resources Planning and Management*, ASCE, 120(6): 962-982.

Todd, D. K. (1956). *Groundwater Hydrology*. 2nd ed., John Wiley & Sons, Inc. New York.

U.S.D.A.-S.C.S. (1964). Hydrology Section 4 Part I, Watershed planning. In: *National Engineering Handbook*. US Department of Agriculture Soil Conservation Service, Washington D.C.

Velázquez P.M., Andreu J. and Sahuquillo A. (2006). Economic optimization of conjunctive use of surface water and groundwater at the basin scale. *J. Water Resources Planning and Management*, ASCE, 132(6), 454-467.

Willies, R. and Yeh, W.W-G. (1987). *Groundwater system planning and management*, Prentice Hall, Eaglewood, Cliff, N.J.

Wrachien De Daniel and Fasso A. Costantino. (2002). Conjunctive use of surface and groundwater: Overview and perspective. *J. Irrigation and Drainage*, ASCE, 51(2), 1-15.

Yan, J. and K.R. Smith. (1994). Simulation of integrated surface water and ground water systems – Model formulation. *Water Resources Bulletin*, 30(5), 879-890.

Yu, W. and Haimes, Y.Y. (1974). Multilevel optimization for conjunctive use surface water and groundwater. *Water Resources Research*, 10(4), 625.

10

An Economy-Environment Integrate Statistic System

Giani Gradinaru
Bucharest Academy of Economic Studies
Romania

1. Introduction

The economy-environment inter-conditioning is so visible that any attempt to fundament this affirmation is useless. The environment furnishes the resources, which represent the nucleus of the supply economic activity. The supply is made for two main reasons: goods and services production and consumption. Both production and consumption create wastes, and these wastes are evacuated in the environment (Fig. 1).

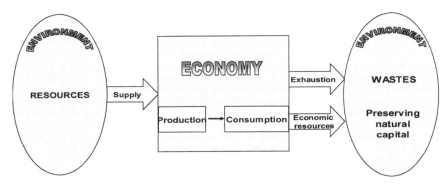

Fig. 1. The economy-environment inter-conditioning

The concept of lasting development doesn't leave any space for a separate discussion on environment economy. The Rio Conference, the 21 Agenda, the Johannesburg summit, scientists' workshops has already established the conceptual basis for creating an environment-economy integrated informational system. No country can be left out because one cannot set boundaries on the environment.

Identifying the information categories relevant for decisions grounding can start the achievement of an informational system concerning lasting development, even before the theoretical and methodological development is fully settled. Thus it can be considered determinant five information categories:

* Highlighting the environment status based on separate environment factors (water, air, soil, biological diversity);

- Emphasizing the environment pressures based on sectors considered pressure sources;
- Estimating the expenses made to avoid pressures;
- Evaluating the size of environment advantages and damages according to the environment pressures;
- Highlighting the standards, which can regulate the pressure.

The first four information categories are so strongly connected to the evaluation problems, that the only problems that may appear are connected to the data collecting methods and the efforts of collecting and processing. On what concerns standards value, there are understandings which count more or less on scientific appreciation, which sustains a higher level of information quality, also generated by the fact that these are decision elements in solving environment problems. An incomplete and uncertain information may influence the consequences of economic activities such as subsequent development.

Achieving the lasting development objectives on a large scale presumes that economic policies are projected according to environment considerations and to the economic functions of the natural resources. For this, the deciding persons need info concerning economic activities and environment status expressed in natural and monetary units. Such information must be built in a manner, which allows an emphasis of the main problem of the lasting development and the inner-generation equity, keeping the environment health for future generations.

The efficiency of the economic reform policies can be evaluated by comparing the traditional synthetic indicators with the ones resulted from integrating the environment data. A simple comparison of these indicators can supply an adequate understanding for introducing the environment parameters in an economic system, reason for that is necessary the use of economic-mathematical modeling.

Because economic policies must be projected in the light of their impact on the environment, the environment policies must take into consideration the economic implications. This integration became nowadays a basic problem in conceiving environment policies, for which the integrated economy-environment indicators can facilitate a coherent wording.

The standard economic indicators, which describe mainly the financial flows in an economy, supply incomplete information concerning the implications of economic activities on the environment. The economic instruments have different possibilities of comparing their results in time and space, but such methods are not developed for environment. The environment informational instruments are usually based on physical parameters, while the economic informational instruments use both physical and value data. As a result, there are significant deficiencies for the quality level of the indicators which must explain the economy-environment inter-dependence, a fact which imposes developing integrated indicators to express the direct connection between the economic activities and environment, in the direction of lasting development request (Fig. 2). In this direction we can define the following priorities:

- The necessity to develop vertical connections between the economic instruments for macro and microeconomic level, respectively between the individual environment indicators (microeconomic level) and the synthesis indicators (macroeconomic level);

- The necessity to develop horizontal relations between the economic and environmental instruments for sector or regional level, respectively including the environment indicators in an economic decisional process;
- The necessity to represent the environment indicators in time dimension, respectively building chronological series for most part of the indicators expressing the economic and environmental performances.

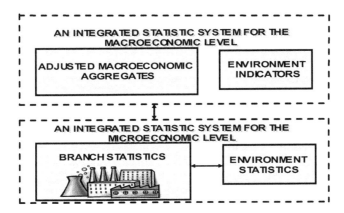

Fig. 2. The economy-environment integration for informational

2. Environment statistic subjects and variables

The statistic base for the economy-environment integrated statistic analysis is complex and perfectible, and this improvement can be achieved gradually. A comprehensive description of the environment requires the integration of a large number of data sources in order to reach a more complete image of the pressure exerted on the environment, of its quality state and of the efforts made to protect the environment. In our country, the present state of these data sources is situated at very low quality level, representing the main obstacle in developing the environment statistic system. The primary data for building the environment statistic-economical analysis system can be improved by developing statistic registers, national accounting revision, doing new statistic researches and improving the existing ones.

Developing the system of environment statistic subjects and variables must be preceded by a clarification of aspect concerning:

- The universal statistic language, the coherence of statistic description being given by the rigorous classification of the statistic subjects, a classification which allows comparison between the information referring to different time periods or different geographical areas (in order to be efficient a statistic language for the economy-environment relation must be systematically developed, so different types of standards to become compatible and establish relations between different information);
- Developing work programs for data gathering and dissemination for subjects such as: emissions, water prevailing and use, waste flow, chemical use, environment protection expenses, available sector statistics for description of environment impact activities etc.;

- Attracting in the environment statistic circuit of those data corresponding to administrative sources, for filling the data fond necessary to comprehensive reflection of the environment problems such as: climate change, air acidification and pollution, exhausting natural resources, exhausting and polluting water resources, urban environment deterioration and waste flow;
- Adopting the European definitions, classifications and unique harmonized naming by the Governmental and non-governmental institutions;
- Projecting and implementing a coherent survey system which would use questionnaires that cover essential domains of the environment statistics;
- Building statistics based on calculus models for those domains, which cannot be informational, covered by statistic surveys or administrative data, such as the case of greenhouse effect gases, chemicals diminishing the ozone layer etc.

When developing the set of environment statistic variables, the present international achievements indicate a preference for combing the environment elements based approach with the one referring to the pressure-status-response and few components from the resource management approach. In Romania, the National Statistic Institute maintains the same conception, a reason for which we would restore a possible draft for developing the environment statistic system of subjects and variables, a draft achieved by combining the three types of approaches mentioned earlier (Fig. 3).

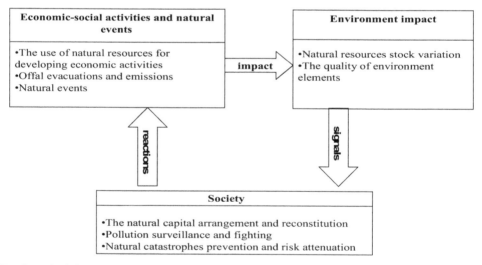

Fig. 3. A draft for developing the environment statistic subjects

2.1 Natural resources use for developing economic activities

Natural resources consumption represents an activity with environment impact generated by the impossibility to restore the consumed resource in a short period. The economic activities with potential environment impact are agriculture, forestry, hunting, fishing, and mining and the extractive industry, energy production and consumption, water use, soil use and landscape transformation.

Agriculture may have environmental incidence by raising the production determined by a growth in cultivated surface or animal effective, either by a growth in agricultural efficiency (Fig. 4.).

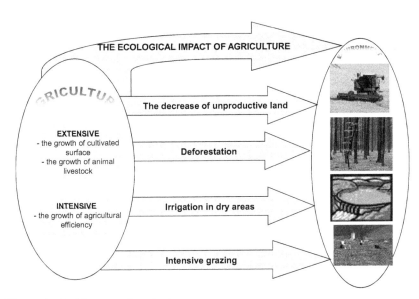

Fig. 4. The ecological impact of agriculture

The statistic subjects which reflect the connection between agriculture and environment refer to a growth of agricultural production (either by extensive agriculture, or intensive) and emphasizing the place of agriculture in the market economy (table 1).

Agriculture	Statistic variables
Extensive agriculture	Cultivated surfaces and the production achieved grouped by types of culture
	The animal effective and their density by animal species
Intensive agriculture	The applied quantity of fertilizers and the fertilized surfaces by types of fertilizer substances used
	The fodder quantities consumed by animals grouped by type of fodder
	The agriculture energy consumption by types of energy
	Agricultural practices based on types of works
The place of agriculture in the market economy	The sales volume based on different types of production (physical and value)
	The inputs volume (physical and value)
	The gross capital formation on types of agricultural exploitation
	The exported volume by different types of products (physical and value)
	The volume destined for self-consumer

Table 1. The statistic variables for highlighting the agriculture-environment relation

The forestry activity has a negative impact on the environment because of forest commercial exploitation, but also a positive impact because of forestation interventions (Fig. 5.).

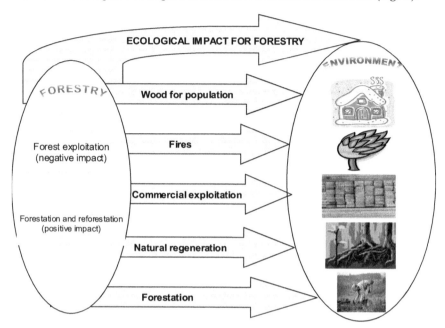

Fig. 5. The ecological impact of forestry

The statistic subjects identified by forestry activity and forest exploitation, connected to environment components consider the commercial exploitation of the forest; wood samples for population, natural loss (fires, diseases, pollution), natural regeneration and forestation (table 2.).

Forestry	Statistic variables
the commercial exploitation of the forest	The exploited timber quantity by types of species
	The primary timber production
	Timber exports without any transformation
natural loss	Losses on types of essences
	The deforestation on types of essences
regeneration and forestation	The annual timber growth on wood species
	Naturally regenerated surfaces
	Forestation surfaces

Table 2. Statistic variables for highlighting the forestry-environment relation

The hunting activity may also influence the environment by destroying fauna habitats as a result of over exploitation or by deliberate destruction of the injurious species (Fig. 6.).

The statistic variables that reflect the connection between hunting and environment refer to the hunting harvest and the economic contribution of this activity (table 3.).

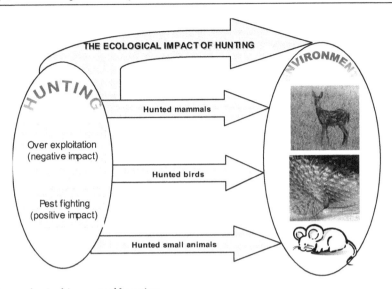

Fig. 6. The ecological impact of hunting

Hunting	Statistic variables
	The hunted mammals effective, on species
	Small animals hunted effective, on species
	Number of hunted birds, on species
	The venison commercial value, on species
The economic Contribution	The income selling specific equipment, hunting permits and the use of tourist infra structure, on types of products
	The export on types of products

Table 3 . The statistic variables for highlighting the hunting-environment relation

The fishing activity can create serious environmental damage by over exploitation and the use of "brutal" fishing methods (Fig. 7.).

In a similar way to the hunting activity, in the case of fishing the main statistic subjects which can be developed refer to the fishing capture and the economic contribution of this activity (table 4.).

The impact of fishing and the extracting industry on the environment can be analyzed by referring to the following statistic subjects, associated to the cycles of mining exploitation (Fig. 8.).

Fishing	Statistic variables
	The fish quantity harvested from the sea, on species
Fish capture	The fish quantity harvested from internal waters, on species
	The fish quantity harvested by sporting fishing, on species
Economic contribution	The export of fish products, on species

Table 4. The statistic variables for highlighting the fishing-environment relation

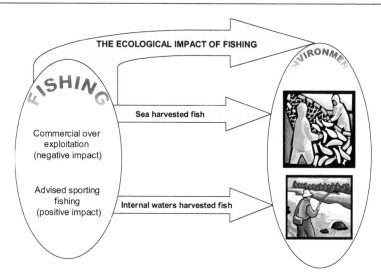

Fig. 7. The ecological impact of fishing

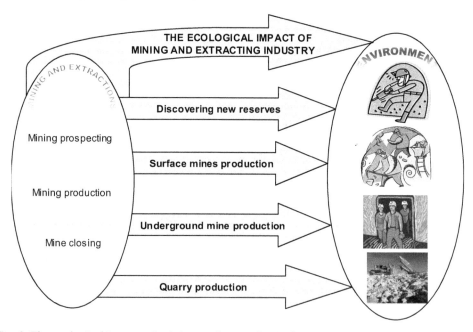

Fig. 8. The ecological impact of mining and extracting industry

The statistic variables, which can be used to analyze the ecological impact of mining and the extracting industry, are presented in table 5.

The energy production and consumption have significant environment impact from the consumer perspective, and also from the pollution perspective. (Fig. 9.).

Mining and extracting industry	Statistic variables
Mining exploitation by mine prospecting	The newly discovered reserves by mineral type
Mining production	The underground mines production by mineral type
	The surface mines production by mineral type
	The quarry production by mineral type
Mine closing	The number of closed mines by mineral type
The role of mineral resources in the economy	The value of mining production by mineral type
	The gross mineral export by mineral type
	The metallurgy consumed minerals by procedure type

Table 5. The statistic variables for highlighting the mining, extracting industry- environment relation

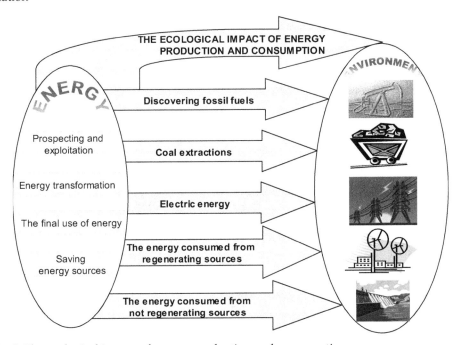

Fig. 9. The ecological impact of energy production and consumption

Table 6 presents the statistic subjects and variables highlighted by the analysis of the impact of energy production and consumption activities on the environment.

The water use is the origin of many environmental problems, especially when the drawings from water bearing formations are made in a rhythm superior to the supply one. In dry areas, the drawing from watercourses may reduce the water quantity at the disposal of downstream users. Another subject of the preoccupation is constituted by the quality of residual water, when it's disposed in water streams, lakes or sea. The statistic subjects and variables, which create the framework of the statistic-economic analysis of water use, are given in table 7.

Energy production and consumption	Statistic variables
Prospecting and exploitation activities	The prospecting number (oil resources, natural gas and coal, discovering other fossil fuel prospecting), by resource type
	The volume of natural gas and oil extraction
	The coal volume extraction
Energy transforming activities	The quantity of fossil fuel used to produce thermal energy, by fuel type
	The electricity production based on fossil fuels, by fuel type
	The electricity production based on classic systems, by types of sources
	The electricity and thermal energy production from unconventional, by sources type
Energy final use activities	The intermediate energy consumption by activity type or industries
	The energy final consumption by activity type
Energy administration	The energy consumption/ habitant and by energy type
	The proportion of energy consumption from renewable or non-renewable sources
	The imported or exported energy by energy type

Table 6. The statistic variables for highlighting the relation: energy production and consumption – environment

Water use	Statistic variables
Water drawing	Water drawn from surface sources, by types of sources
	Water drawn from underground sources
Water use	The water quantity used in agriculture
	The water quantity used in industry, by activity type
	The water quantity used for energy production
	The water quantity consumed by households

Table 7. The statistic subjects and variables for the water use analysis

The set of statistic subjects and variables which can be used in the soil use and landscape transformation analysis, as distinct aspects of the natural resources use with the purpose of economic activity development, is given in table 8.

Soil use and landscape transformation	Statistic variables
Changing the destination	Changes in soil use between activity sectors, by soil use form
	Changes in soil use inside the economic sector, by soil use form
Environment reorganization	The transport network by transport type
	Hydrological structuring by creating dams, accumulation lakes, canals
	Creating residential and industrial areas
	Achieving substructures for: mining, forest exploitation and commerce

Table 8. The statistic subjects and variables specific to the economic activity which imply soil use and landscape transformation

2.2 Waste emission and exhaustion

The statistic analysis of waste emission and exhaustion resulted from economic- social activity consider the air polluting substances emission, the exhaustion of polluting substances in water and generating wastes. The statistic variables used in such analysis are presented in table 9.

Waste emission and exhaustion	Statistic variables
Polluting substances air emission	The emitted quantity of polluting substances (nitrogen oxides, carbon oxides, ammonia, organic and inorganic compounds, heavy metals, suspension powders) by type of polluting agents and activities
Polluting substances water exhaustion	The volume of residual water from the public sewage system, by type of polluting agents and hydrographic basins
	The volume of industrial residual water, by type of polluting agents and hydrographic basins
	The quantity of polluting substances resulted from agricultural practices (diffuse pollution due to agriculture)
	The quantity of polluting substances in the rain (diffuse pollution due to acid rainfall)
The amount of wastes	Soil, waters, air waste evacuations
	Other variables can be developed in concordance with the waste classification

Table 9. The statistic variables for highlighting waste emission and exhaustion

2.3 Natural events

These statistic subjects contain variables referring to natural phenomena, which can affect human population production, consumption and welfare. There is a synergetic effect between the natural phenomena and the impact of human activity on the environment. For example, a poor land use during drought may stimulate a deserting phenomenon; building human settlements in vulnerable or seismic areas cause destruction and human lives loss (Fig. 10.).

The variables, which come out of, figure 10 describes the size and intensity of natural events, classified by meteorological geological or biological origin (table 10.).

Natural events	Statistic variables
Meteorological risks	The rainfall volume and its variation according to the average
	The temperature and its variation according to the average temperature
Geological risks	The number of earthquakes
	The landslides number
Biological risks	The insect infested surfaces
	The number of epidemics

Table 10. The specific statistic variables for describing the natural events

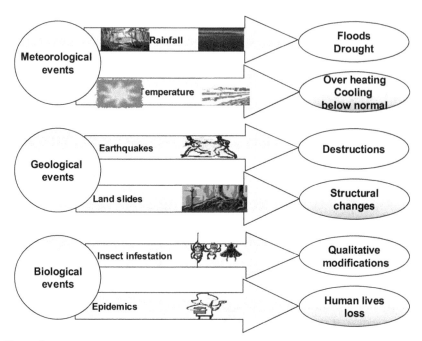

Fig. 10. Natural events

2.4 The natural resources stock variation

The statistics subjects included in this category devolve from the informational system, which considers the lasting resource management. The emphasis is put on growth or reduction of biological resources stock and of cyclic resources (water, soil and minerals). From the environment's point of view, a reduction of the biological resources stock happens when the exploitation exceeds the natural regeneration rhythm (table 11.).

Environment elements	Statistic variables
Soil/subsoil	The net variation of land surfaces
	The net variation of annual cultures biological mass
	The livestock net variation
	The net variation of forests
	The variation of forest biological mass
	The losses of productive soil due to land use changes
	The losses of productive soil due to corrosion
	The initial reserve of mineral resources by mineral type
	The annual production of mineral reserves by mineral type
Water	The variation of the fish population
	The level variation for surface water
	The modification of the average flow of water streams
	The variation of the lakes stocking capacity

Table 11. The statistic variables for the analysis of natural resources stock variation

2.5 The environment elements quality

The level of air, soil and water quality is generally compared to standard quality norms. The statistic variables may express in terms of variation and frequency, the exceeding of admitted standards and the quality degree of each environment element (table 12.). The air quality is mainly determined by environmental pollutant concentrations. In this direction, in order to control the polluting chain and fight its effects must be used in all of the pollutant production and transmission stages the same concepts, definitions and classifications. The notion of water quality is more complex than the one of air quality, because quality parameters depend besides the pollution agents on the water uses. Establishing the statistic variables referring to water quality is necessary because of the preoccupations concerning the contamination of the hydrographic network, due to pollutant evacuations from industrial, agricultural and human settlements units. Soil quality influences the productivity of biological systems, and their degradation reduces the production of biological mass and the capacity to produce services. Natural factors, but also agriculture practices determine the variation in soil quality. In this manner, soil degradation can be connected with wrong work and use practices for agricultural crops, but also with land improvement activities and excessive use of agrochemical products. Also, a certain state of soil quality contributes to acid deposits resulted from atmospheric pollution, the humus loss as a result of chemical fertilizers applications, soil compaction due to agricultural machines use.

Environment element	Statistic variables
Air	The maximum effective concentration of pollutants by type of pollutant agents
	The average monthly/annual pollutant concentration by type of polluting agents
	The frequency of exceeding the maximum admitted concentration
Water	The physical and chemical properties of used waters
	The chemical substances concentrations
	The organic matter concentrations expressed by COD (Chemical oxygen demand – the oxygen quantity took from the water organic matter, used as a measurement unit for organic matter in domestic waters)
	The river length by quality classes
Soil	Deserted surface
	Eroded surface
	Toxic substances contaminated surface
	The acid deposits surface
	The surface of irrigation degraded soil

Table 12. The statistic variables for the emphasis of environment elements quality

2.6 The arrangement and reconstitution of natural capital

The improvement of resource consumption in the classical meaning, considers an economic productivity maximization of natural resources for the growth of production units which use these resources, such as: agriculture, forest exploitation, fishing and the extracting industries. A statistic expression for the arrangement and natural capital reconstitution can be achieved with the help of following statistic variables highlighted in table 13.

The arrangement and reconstitution of natural capital	Statistic variables
Nature protection and conservation	The national parks network
	Protected areas
	Protected fauna
	Public expenses for arranging and restoring natural resources
	The personnel engaged in protecting and conserving nature
Reconstitution of degraded environment	Restored agricultural land
	Trees cultivated land
	Protected species of flora and fauna

Table 13. The statistic variables for describing the arrangement and reconstitution of natural capital

2.7 Pollution supervision and control

There are more categories of statistic variables, which can describe the activities of pollution supervision and control (table 14.).

Pollution supervision and control	Statistic variables
Pollution research and environment supervision	The number of researches concerning pollution
	The number of air or water quality supervision stations
Environment restoring and pollution fighting	Pollution fight by pollutant type and ecosystem
	Restoring operation by ecosystem and pollutant type
The means to fight pollution	The number of water treating stations by type of treatment and hydrographic basins
	The mud quantity evacuated by hydrographic basin type
	The quantity of dangerous waste treated
	Public funds allocation for the enterprises pollution fight
Actions started by enterprises	The volume of cleaned residual water
	The necessary cost for managing dangerous waste
	Waste recycling
	Investments for environment protection techniques
	The costs engaged for producing consumer goods which don't endanger the environment
The households' reactions	The modifications in the expenses structure
	Households' waste recycling
	Buying products with low environment impact
	The consumption modalities by type of consumer (Choosing lead free gas, using paper wrappings, choosing the size of the vehicle, etc.)
	Recycling materials
	The population behavior on what concerns the participation to the recycling process

Table 14. The statistic variables for the description of the pollution supervision and control

2.8 Preventing natural catastrophes an risk attenuation

In front of natural forces, the reaction may be scientific, technical, biological, administrative and humanitarian. The variables presented in this category (table 15.) describe the measures took in order to prevent floods, the operation of catastrophes surveillance and foreseeing, emergency measures for reducing their effects (meaning evacuating the population, etc.).

Types of activities	Statistic variables
Prevention	The number of researches by activity type
	The physical substructure for natural catastrophe protection
	Administrative regulations by regulation type
Control	Biological activities

Table 15. The statistic variables for emphasizing the preventing actions for natural catastrophes and risk attenuation

3. Conclusion

The economy-environment integrated statistic analysis can be applied in different stages of the decisional process, such as: identifying environment priorities, identifying the pressure points, projecting environment policy, evaluating the policy's effects. The data can be used to monitor the effects of environment policies in the terms of public and private commercial activities, as well as in terms of positive or negative sector effects induced by different sectors.

4. Acknowledgment

This research was supported by Human Recourses Program, Grant no. TE_336/2010, (agreement no. 45/03.08.2010), financed by National University Research Council (CNCSIS-UEFISCDI).

5. References

Bran, F. (2002). *Componenta ecologică a deciziilor de dezvoltare economică*, ASE Printing House, Bucharest.

Bryant, D., Nielsen, D., Tangley, L. (1997), *The last frontier forests: ecosystems and economies on the edge*, World Resources Institute, Washington DC.

Cox, S., Searle, B. (2009), *The sate of ecosystem services*, The Bridgespan Group.

Ehrlich, P.R. (2002), Human Natures, Nature Conservation, and Environmental Ethics, *BioScience*, vol.52, nr.1, pp. 31-43.

Glaser, M. (2006), The social dimension in ecosystem management: strengths and weaknesses of human-nature mind maps, *Human Ecology Review*, vol.13, nr.2, pp. 122-142.

Grădinaru, G. (2008). *Tehnici de analiză statistică a beneficiilor de mediu*. ASE Printing House, Bucharest.

Grădinaru, G. (2004). *The bases of Environment Statistics*. ASE Printing House, Bucharest.

Grădinaru, G., Colibabă, D., Voineagu, V. (2003). *Quantitative Methods for Environment Data Analyses*. ASE Printing House, Bucharest.

Hannah, L., Carr, J.L., Lankerani, A. (1994), Human Disturbance and Natural Habitat: a Biome Level Analysis of Global Data Set, *Conservation International*, Washington DC.

Hezri, A.A., Dovers, S.R., (2006). Sustainability indicators, policy and governance. Issues for ecological economics, *Ecological Economics*, 60, pp. 86-99.

Lackey, R.T. (1998), Ecosystem management: paradigms and prattle, people and prizes, *Renewable Resources Journal*, vol.16, nr.1, pp. 8-13.

Layke, C. (2009), *Measuring nature's benefits: a preliminary roadmap for improving ecosystem service indicators*, World Resources Institute Working Paper, Washington, D.C.

MEA (2005), *Millennium ecosystem assessment. Ecosystems and human well-being: scenarios 2*, Washington DC, IslandPress.

Negrei, C. (2004). *Economia și politica mediului.* ASE Publishing House, Bucharest.

OECD (1993), *OECD core set of indicators for environmental performance reviews*, Synthesis report by the group on the state of the environment, ENV/EPOC/GEP (93)5/ADD, Paris.

OECD (1997), *OECD core set of environmental indicators*, Biodiversity and Landscape – draft working paper, Group on the State of the Environment, ENV/EPOC/SE (96)13/REV1, Paris.

OECD (1998), *Agriculture and biodiversity*, OECD workshop on agri-environmental indicators, COM/AGR/CA/ENV/EPOC (98)79, Paris.

OECD (1999), *Environmental indicators for agriculture: methods and results – the stocktaking report greenhouse gases, biodiversity, wildlife habitats*, COM/AGR/CA/ENV/EPOC (99)82, OECD, Paris.

OECD Proceedings (2000). "*Frameworks to measure sustainable development*, An OECD Expert Workshop.

11

Sustainable Performance and Environmental Health in Brazil: How to Understand and Measure!

Márcio Ricardo Costa dos Santos
Universidade Federal Fluminense (ASPI/UFF)
Academia de Medicina Veterinária (AMVERJ)
Instituto Camboinhas Ambiental (ICA-Nit)
Brazil

1. Introduction

Living together of people in harmony with nature is fundamental both for social development and for improving the quality of life and must be evidenced primarily by professionals involved in agricultural and health sciences. However, since the beginning of colonization of Brazil our natural resources have been exploited as predatory, and just consider, as an example, the Atlantic forest, one of the most valuable ecosystems of Brazil, which currently has only 7.3% of their original coverage. Our settlers-users enjoyed nature with careless posture and it has been almost totally destroyed. So if undertaken by centuries, through the industrial revolution, and expanding biotechnology for agricultural production, it has been consolidated in diverse environmental problems. The existence of these problems, in all regions of the country, requires the development and deployment of environmental educational programs, important in attempting to reverse or minimize this framework, to become an odd question and impregnate the consciousness of ecologically correct on all levels of our society.

Applications of technologies innovations contribute always more for global development; from the economical, social and environmental point of view, to produce more with respect to the living beings, and they are becoming hard to observe. The enterprise productive activities demand attention for research and the application of environmental evaluation systems, before the magnifying of the use of biotechnology products, because of the gradual focus changing to the green chemistry, biopharmaceuticals and bioenergy. Certification and rastreability methods gain importance on this process, with the establishment of multiple or mutually exclusive objectives between the social agents. Among the possible alternatives to carry out assessments of socio-environmental performance of urban and rural activities, the use of ecological and social indicators of sustainability and health has been a method of choice. The indicators should be organized in impact assessment systems that may span increasing levels of complexity and goal requirements for the environmental health management (Santos & Rodrigues, 2008; Santos, 2010).

2. How to understand

2.1 The goal of the Brazilian Agenda 21

International and national commitments undertaken since 1992 at the United Nations Conference on Environment and Development (Eco-92), in Rio de Janeiro, already showed that environmental conservation and maintenance of health should walk together with the development model to be exercised thereafter, to generate prosperity. Challenges for the new century, we highlight the "Global Development Agenda and Sustainability - Agenda 21" with empowerment and engagement organizations, facilitating access to information and financial resources, and defraying ongoing activities on the global aspects of specific impacts. The 175 nations attending the Conference Eco-92 approved and signed the Global Agenda 21 and undertook to comply with its terms. It represents the basis for the continuation of life on the planet and the construction of a new development model, capable of achieving social justice and promotes decent conditions of life for present and future generations, without aggression to the environment or depletion of available resources.

One has to prioritize education and health, with the creation of public policies, human and animal rights for sustainable development. There are challenges apply to a society, which must act on themes such as water and sanitation management, assessment of climate change, income generation, poverty eradication and environmental justice. The plan was prepared by all countries, in dialogues and negotiations which had the participation of representatives of all layers of society present at the event. These recommendations also can steer Agenda and establish ways to live better and with good living conditions for all; they are ideas and principles which aim to create a sustainable world. It was hoped that the participatory process of commitments, actions and goals, to transform the development on the basis of the principles of life (the recommendations of Agenda 21), were adopted before the 21st century, i.e. between Eco-92 and the end of the year 2000, what did not happen everywhere in the world.

The construction of the Brazilian Agenda 21, the Commission of Sustainable Development Policies and National Agenda 21 (CPDS) aimed to introduce the concept of sustainability, to qualify the potentialities and vulnerabilities in Brazil in the international framework. The Brazilian Agenda 21 was released on July 16th 2002, on the site (http://www.mma.gov.br/sitio/ index.php?ido=conteudo. monta& & idEstrutura = 6 idConteudo = 908) in which refers to its content and it is possible to realize the nature of the process of strategic planning and participatory, to be implemented in stages. It is a widerange approach to a product of consensus among the different sectors of Brazilian society. At launch, six thematic areas were covered and named: sustainable agriculture, sustainable cities, infrastructure and regional integration, management of natural resources, reducing social inequalities, and science and technology for sustainable development. All this material is published in two documents as a result of discussions throughout the national territory, namely: the result of the national consultation (http://www.mma.gov.br/estruturas/ agenda21//consulta2edicao. zip files) and the priority actions (http://www.mma.gov.br/ estruturas/agenda21/_ files/acoes2 edicao.zip).

The goal of the Brazilian Agenda 21 deals with the promotion of sustainable agriculture. It reveals the desire of society to enjoy healthier products, with high levels of food safety, environmental, and it involves the technological innovations of the productive sector, which provide new methods applicable to agriculture. The solutions considered "sustainable" often require specific ecosystems and ecological knowledge, the large-scale demand that, initially, a work of education and awareness to preserve natural resources with sustainable performance. However, there is a situation that does not change so rapidly in population without basic knowledge of environmental health and what prevails is just the pace of innovation. The strength of the innovative trends regarding the different types of innovation in the industry generates conflicts between food production and preservation of ecosystems. Regulatory modalities of these conflicts become decisive so that there is peace in the rural and urban food productive sector.

Let's consider for example the Atlantic forest, which is very low, very fragmented in its original area and has currently only about 102.000 km to house all kinds of living things and keeping them, including fed! It is the second most threatened biome in extinction of the world; it loses only virtually extinct forests of the island of Madagascar in the African coast. Even so, the Atlantic forest biome presents an estimate of shelter around 1.6 million species of animals, including insects. There are cataloged 261 species of mammals, of which seventy three that only exist there and nowhere else on the planet Earth, are endemic in this biome, against 353 species catalogued in the Amazon, while Amazon biome is four times larger than the original area occupied by the Atlantic forest. There are 600 species of birds, of which 181 also are endemic of this forest. There are 280 species of amphibians, and 253 of this are biome endemic species! Reptiles reach 200 species, of which sixty are endemic. It is estimated that the biome is home to more than twenty thousand species of plants, of which eight thousand are endemic. It is the richest world's forest in diversity of trees and can provide, as in the South of Bahia, up to 454 different species per hectare!

All these figures actually represent an enormous wealth of biological diversity; is a greatness, which causes all of us proud to be living creatures in the vast biodiversity of an important Brazilian biome. But there are about 120 million people who live in the area of the Atlantic forest, which means that the quality of life of approximately 70% of the Brazilian population depends on the preservation of the remnants of the Atlantic forest biome! In this forest is the abstraction of water that keeps the springs and sources and that regulates the flow of fountains, in addition to providing conditions for cities and society may have supply at all levels of environmental health: it consists in regulating the climate, temperature, humidity, precipitation processes promotes rainy region, ensures soil fertility, protects the slopes of hills, slopes and creates conditions for reproduction and perpetuation of wild and domestic species in the region. Anyway, it allows the existence of life!

2.2 Sustainable performance a question of mentality

To have a sustainable performance is a question of mentality and it means changing your habits for better and having a healthier life (Figure 1). The biggest challenge, according to the authors: the population will put into practice in their homes and also in their cars, with near zero cost, important actions searched. Some of the measures envisaged are: to use clean energy, refrigerators and cold lamps with certification, which consume less energy; dry clothes outdoors instead of using machines; clean and adjust the air-conditioning and heating; choose LCD televisions instead of plasma apparatus; turn off appliances from stand-by; save water, etc. In automobiles, the measures involve the owner to exchange regular engine oil, systematic verification and control of emissions, as well as maintaining the recommended pressure in the tires. In addition to these actions, the driver shall drive sparingly in speed, without abrupt maneuvers, and avoid unnecessary braking, which increase fuel consumption. To complete, scientists encourage attitudes of courtesy and partnership between citizens, and the use of a single vehicle by several known people to travel with the same goal and to the same places.

Fig. 1. A beautiful view from the top of Rio de Janeiro skyscrapers: how beautiful are the socio-environmental health and the sustainability down there?

For example, people need to think carefully that agriculture needs to adapt to the new rules of garbage treatment. Decree lays down rules for waste disposal and reutilization from the industry, including refrigerators, sugar mills and alcohol and packaging companies: The agricultural sector should adapt to new rules for the treatment and disposal of garbage. Decree Nº. 7.404, published on January 23rd 2010, regulates the National Solid Waste Policy and establishes the standards for selective collection and refund of the productive sector solid waste for reuse or other environmentally appropriate. The legislation includes, for example, the procedures for the manufacture of pet food from bovine bone and the use of biomass as residue of sugar cane, for energy production. One must comply with manufacturers, distributors and sellers of used packaging or other wastes, involving products such as pesticides, batteries, tires, lubricating oils, light bulbs and electronics. The determination is valid for companies which have concluded agreements with the public sector for the deployment of shared responsibility through product life cycle (sectored agreement).

The legislation also strengthens the recoil, and reusing packaging of agrochemicals and the treatment of products seized and waste produced in ports, airports and border procedures already provided for by law. "The Decree represents a breakthrough in the treatment of waste in the country and ensures the use of by-products and residues of animal and vegetable regulated industries by unique attention to livestock production sanity (Suasa)". The measure applies to the processing of plastic containers, cardboard and disposable plastic bottles; they now should be reused or recycled and can no longer be destined for the landfills (Figure 2). According to the Ministry of Agriculture the law establishes and reinforces the concepts of non-reporting, reutilization, recycling, reuse, treatment and energy recovery from waste discarded by the productive sector. The law also provides for the replacement of garbage dumps by landfill for waste, the creation of municipal plans, state and federal for waste management and encouraging cooperative funding lines, which should assist the selective collection and reverse logistics. The regulation requires that the procedure of urban collection at least separate recyclables and humid.

Fig. 2. Population depends on the preservation of the Atlantic forest biome remnants and disposable plastic bottles should not be destined for the rivers!

The International Organization for Standardization's (ISO) new guidelines are internationally accepted standards, on social responsibility for all organizations, public, private or third sector, and the World Society for the Protection of Animals (WSPA) celebrates the inclusion of animal welfare in the newly launched ISO Standard 26000. This published inclusion of animal welfare in ISO 26000 marks a historical first step, as it will encourage a large number of entities to acknowledge the importance of animal welfare while pursuing certification as per ISO standards. Their WSPA international programs manager believes that this is a landmark, because it states unequivocally that animal welfare matters to all. "Our actions impact animals in countless ways and accordingly, we have a responsibility to ensure their welfare is respected, be it as a company, school, municipality, church, university, ministry or in any other form we organize ourselves." The standard, although voluntary, is highly significant as it states what organizations need to do to operate in a socially responsible manner, stipulating that the welfare of animals used economically, or in any other way, must be taken into consideration.

Most notably, the text of ISO 26000 states that organizations should aim at: "respecting the welfare of animals, when affecting their lives and existence, including by providing decent conditions for keeping, breeding, producing, transporting and using animals." The standards also make specific mention of the physical and psychological wellbeing of animals in several chapters, not restricted to the section relating to the environment. The need to take animal welfare into consideration in an organization's business practices has also been integrated into actions governing ethical behavior, consumer issues and community involvement and development, specifically in wealth and income creation.

The inclusion of animal welfare in these standards is a result of a very hard working from sustainable citizens, i.e. WSPA, together with Dutch consumers association and others, which has been presented in the working group discussing the development of ISO 26000. Discussions on ISO 26000 began in 2005 and the working group which ISO formed to develop the standard was the largest in the Swiss-based organization's history, involving around 400 experts from 99 countries. After an initial focus on corporate responsibility, it was decided to open the standard up to include all types of organizations. "So far, because animal welfare was not part of the ISO benchmark, organizations could claim to take their social responsibility seriously, despite overlooking the interests of the animals affected by their business practices," said WSPA's international programs manager. We all hope that other organizations responsible for issuing guidelines or standards – such as the IFC or OECD – will follow the ISO example, giving animal welfare its rightful place as a critical aspect of social responsibility. The ISO Working Group agreed that it made sense to include animal welfare as an element of social responsibility, because the specific issue of animal welfare is a relatively new subject in ISO standards. Bringing together so many experts from different stakeholder interests to debate this new standard has helped to ensure that the final consensus represents a depth and breadth of input on social responsibility as a whole.

2.3 The role of health professionals – Veterinarians[1]

Here comes the need for reflection on the role of the Doctors Veterinary Medicine, especially one that deals with the formation of new professionals and that has often taken a practice uncoupled from research and continued learning. The veterinarian who performs an educational work acquires its political potential in as much as it promotes technical skills, and values of the scientific information and it can involves students and producers in the process of professional development; it must generate a space for discussion, in search of solutions to the problems and the limiting situations, with paradigm shift to sustainable. It is important to consider that the interest groups have particular ways of understanding reality, of interpreting the alternatives and commit themselves with the proposals, but they must participate in decisions about what to put in place to better involvement and development of the profession: forming brains is more profitable and significant long-term than selling land, logic that reverses the flow of current use of technologies developed abroad and promotes intellectual property country for export. Universities and research centers or Brazilians institutes are fundamental to the development of technologies, innovation, quality assurance and environmental management, and also for finding alternatives which benefit populations in general.

The extension projects (Figures 3 & 4) are very important in this context, to act on the interface producer/production, processing and warehouse/distribution/social development. The extension teaches and guides the population in the deployment of development projects, being a channel to lead the technology involved in the production chain of those who are outside of research cores. Prospects and guidelines for the sustainable development should be addressing the implementation of measures integrating society with these institutions; that also spreads the results of work undertaken in obtaining

---

[1] In honor of the World Veterinary Year – 2011; the 101st birthday of the first Veterinary School creation in Rio de Janeiro – Brazil, and the 250th birthday of the first Veterinary School creation in Lyon - France (*la première École Vétérinaire au monde*)!

Fig. 3. Marine farms in Angra dos Reis, RJ: cultivation technologies in hatcheries.

Fig. 4. Students visiting mollusks cultivation in hatcheries at the marine farms.

information about the economical situation, social and environmental area and promoting the necessary interventions, both through technical assistance and policy decisions inherent promotion of development, based on sustainable management of technological innovation.

We believe that the veterinarian has a key role in raising awareness and regulation of this area and in these involved industries, including the agribusiness. For example, only the correct usage of pastures in degraded areas in Brazil could avoid the emission of 80 millions tons of CO_2 in equivalence a year. This value is bigger than the Brazilian industrial emissions that according to the latest research in 2005, was 78 millions of tons in equivalence. The numbers were calculated by Icone, a partner of Getulio Vargas Foundation (FGV) in a launched report about the emissions in the agricultural sector in 2010 (Beef Point). The need of the veterinarian in rural areas is also growing, not only as a health professional or in the control of epidemics, but mainly for the environment. This concern with awareness encompasses all areas occupied by these workers, necessitating even social, cultural, political and economic interaction.

However, many newly trained, professionals do not consider themselves qualified with the requirement of market to offer creative ideas on how to assist and practicing sustainability in its profession; and such due to the importance of a trader in so many areas, their training is fundamental to it can express all their capacity. According to our survey, the opinions of veterinarians are quite different. Only a small portion defends that veterinarians can operate easily on a multidisciplinary and only means baggage graduation, while most think that the same learn more about education and social responsibility, not only in relation to the issue of sustainability, but also by the degree of relevance which this professional has in public health and health animal. This indicates that there is a need for greater attention to the collective health, taking into consideration social, cultural and economical aspects and will have a higher degree of commitment to the health of the population.

The lack of interaction between veterinary medicine preventive and public health with the humanities and social sciences is a reflection of the lack of interaction of courses with other areas of knowledge. Veterinary medicine in the 1980s was ranked as one of agronomical sciences in Brazil and since then, conversely, was being simplified in its most basic principles that could express the medical activity with intervention in clinic, in human and animal health. Similarly, studies concerning the ecology and the environment, that in whole form a single, indivisible biosphere and impossible to be understood through pieces were underestimated. This absence of a systemic and holistic look led to the formation of professionals more directed to the production of animal origin foodstuffs and their derivatives, to take account of the accelerated process of industrialization of western countries and to the dictates of an uneducated and rambling consumer market, which prioritizes the immediacy in detriment of the ecosystems on which was created. Something was broken, complex production systems, adapting social, cultural and environmental areas were disrupted before being examined as to their sustainable efficiency, and in many cases in which progress has brought misery, we didn't know how to revert any more and neither are we able to conjure up something that may replace it. This is the true complexity! We have a sequence of events which escape our cognitive capacity, about which we talk and understand what we've done and what we can do to avoid aggravating more distortions generated by our ignorance, immediacy or disengaging with nature and the future of us all.

In the new century are expected from health professionals, educators and agronomical sciences, full of knowledge of all the intricacies surrounding the work, performed worldwide with Agenda 21, is either global, national, state or municipal, but that features the construction of a healthier world at present and that extends into the future, getting better for people and animals. Sustainable performance in veterinary medicine comprises a participatory process that can: a) provide awareness of involved veterinarians, with changes of attitudes in the quest for integration between the different sectors of society, in line with the principles of sustainability, to consolidate a culture of professional development and biotechnological innovation; b) apply to students, clients and other members of the community the full exercise of social responsibility, according to their professional competence; c) contribute to improving human relationship amongst themselves, with participants in society and with the animals, to exercise the ethics, the well-being and health habits, for personal hygiene and food; d) promote interactive practices, values and discuss with colleagues, students, clients and public participant, to teach new ways and opportunities for insertion into a quality market, with the adoption of appropriate technologies to environmental management, either in the agribusiness as a whole, whether in rural or urban activities too; e) generate new money in qualified and well-paid positions, through consultancies and expert services run by veterinarians involved with sustainable business management.

A new methodological approach with the development of perception of environment-dependent health can be a key instrument for awareness and changing values and attitudes in rural and urban area. Educational work effectively targeted towards the human being, but that reaches the animals, the environment and meets the demand of processes to a Veterinary of the new millennium, in which the exercise of the profession and the production of food become a premise of sustainability. Health professionals can thus contribute to building a sustainable society, working in training multipliers in the various sectors of the communities where they operate and develop the perception of collective learning and knowledge of environmental reality of each of its ecosystems. In order to exercise awareness of society about the importance of the environment to health, for the production of safe food and for a sustainable future should obey this methodology: an educational mass process exercised in their daily deals with the capture of multipliers in different sectors of their communities, including veterinarians, in public services, private and civil society. The proposed method should result in the compromise of each citizen to perform simple activities and collective interest around improving the quality of the environment and food. It may involve orientation and practices that stimulate cognitive affective capacities, and psychomotor reading about the diversity and complexity of landscape. Environmental safety can meet the issue of agriculture with the analysis of environmental impacts caused and emphasize its importance to human existence in rural areas, including the urban population, with an emphasis on reflection on the issues of strengthening of family agriculture and agro ecology.

The evaluation of health education should assist in continual improvement of veterinary medical education on the health of the population related cultural menu and biotechnology innovation processes, with the establishment of conditions for continuity and permanence of sustainable populations as a result of improvements in environmental quality. The perception of the elements of the environment will be facilitated to the extent that we enter as part of nature, interacting with it as individuals and also in social groups. There is no need to restructure, but to innovate in the methodology, in order to meet demands for an educational process effectively, with contemporary calls like "Sustainable Business" to stimulate partnerships and implement projects. New diagnostic techniques, impact assessment and environmental management not only enable a thematic for agribusiness, but can also promote the execution of simple experiments that help the study of basic and significant issues such as the possession charge of pets in urban or rural environment, micro-organisms important in agro-environment and health impacts of risk by pesticides on aquatic organisms.

For example, fines for violating Canada's Health of Animals Act have more than doubled the Canadian Food Inspection Agency reports. In a statement, CFIA said the Government of Canada is cracking down on those who mistreat and improperly transport livestock with the first increase in fines in over ten years. Stressed animals are generally considered to be more susceptible to disease and infections, making humane treatment an important food safety consideration. Violators now will face fines of up to $10,000, up from the previous maximum of $4,000 for convictions under the Health of Animals Act (Food safety news / Meat Trade News). Veterinary professionals should continually work towards improving animal health and welfare by way of available tools, such as advocacy, scientific research and lobbying to enact appropriate legislation. It is the responsibility of humans to ensure animal welfare and include steps to achieve proper housing, nutrition, disease prevention, humane and responsible care and humane handling of animals.

3. How to measure

3.1 Original research papers and literature

Independently of the agribusiness branch, the certification and rastreability methods gain importance on this process, with the establishment of multiple or mutually exclusive objectives between the social agents. The application of different scales of space and secular measures in the delimitation of the impacts, and the use of not standardized criteria for evaluations are some of the problems which many times contribute for the failure of the studies on environmental impact of new used technologies (Andrioli & Tellarini, 2000).

Among the possible alternatives to carry out assessments of socio-environmental performance of agribusiness activities, the use of ecological and social indicators of sustainability has been a method of choice (Girardin et al., 1999; Bisset, 2000; Santos, 2006). The indicators should be organized in impact assessment systems that may span increasing levels of complexity and goal requirements for the environmental management (Rodrigues et al., 2006; Payraudeau et al., 2004).

The Brazilian Agricultural Research Agency (Embrapa – www.cnpma.embrapa.br/) has proposed a system for environmental impact assessment of agricultural technology innovations (Ambitec-Agro) for the appraisal of research projects and technology innovations in the institutional context of research and development (Monteiro & Rodrigues, 2006; Rodrigues et al., 2006), and in attendance to a demand of the "Inter-American Institute for Cooperation on Agriculture – Cooperative Program for the Agricultural Technological Development of the South Cone" (IICA – PROCISUR, 2006 – www.procisur.org.uy). Another platform from the Ambitec-Agro has been proposed, integrating all environmental and social indicators, to implement the natural resources and biodiversity management in Uruguay, known as Responsible Production Project (Rodrigues & Viñas, 2007).

A derived impact assessment system organized toward eco-certification of agribusiness activities was applied with excellent results to the environmental assessment practice of the aquaculture activities in Rio de Janeiro State and presented in May 2006 (Santos, 2006). Our original paper worked out a sustainability evaluation system of best sea scallops (pectens) production practices and its applications as a project appraisal for the technology innovation management methodology in the Pectens *In vitro* Fertilization Laboratory, to attend the Ilha Grande Bay Marine Farming Program from the Institute for Eco-Development of the Ilha Grande Bay in Rio de Janeiro State (Santos & Zaganelli, 2007). The system comprises a set of weighing matrices organized for the integrated assessment of socio-environmental indicators, including modules focused on the environmental impact assessment of sea scallops production, or similar agribusiness activities, with a specific module for social impact assessment. Another similar platform has been described recently by the author, integrating all environmental and social indicators, to implement the natural resources, biodiversity management and productive enterprises at the Camboinhas region in Niterói, RJ, but is still being published; the study worked out a sustainability evaluation system with 32 socio-environmental health indicators for best production practices and its applications as a project appraisal for different urban and rural activities, with a specific module for health and social impact assessment. The calculated impact indices facilitate to making socio-environmentally sound decision and allow the delineation of recommendations for performance improvements, as well as selection of best cases for benchmarking purposes (Santos, 2010).

Keywords: sustainable performance, environmental health management, sustainability indicators, impact assessment.

3.2 Research methods

An impact assessment system organized toward eco-certification of agribusiness activities was presented in 2006 and applied with excellent results to the environmental assessment practice of the aquaculture activities in Rio de Janeiro State (Santos, 2006, Santos & Zaganelli, 2007). This paper presents the practical impact assessment method for the adoption of technology innovations in the Pectens *In vitro* Fertilization Laboratory (FIV-Lab) from the Institute for Eco-Development of the Ilha Grande Bay (IEDBIG), at the Jacuacanga district, Angra dos Reis, Rio de Janeiro State, Brazil (Figure 5).

Fig. 5. The Larviculture and Oceanographic Research Laboratory from IEDBIG allow the reproduction control of fish spat, spawn and fry and research to increase aquaculture productivity in the region.

To fulfill the system framework requirements, focused on reproductive and productive enterprises, field visits and interviews supported by a structured questionnaire, with the laboratory executive director and farmers were carried out. The indicators related to soil and water quality were obtained by laboratory analysis at UFF, and the other related to historical and administrative knowledge of the director and farmers were obtained by local interviews.

The sustainability evaluation system of best pectens production practices (Best-SES) consists of a 24 integrated indicators set (Figure 6), spanning two dimensions namely landscape ecology and socio-environmental quality. These dimensions are integrated with seven essential aspects to encompass the IVF-Lab and the sea scallops, known as "Coquilles Saint Jacques" (*Nodipecten nodosus*, Linnaeus 1758) marine farms production, within the local environmental market settings. The indicators weighting matrices are constructed in a MS-Excel® platform to translate variables and attributes into environmental impact indices, expressed graphically and related to a utility function of environmental performance benchmarks, derived from sensitivity and probability tests, case-by-case for each indicator. The Best-SES was formulated to evaluate the sustainability indicators, considering its benchmark compliance values, validated in contrasting situations representative of the establishments addressed spectrum, since the implantation of the technological innovations (2006) until the applied Best-SES (2008), according to the previous methodology presented by the authors (Rodrigues et al, 2006; Santos, 2006).

The qualitative research, type case study, followed a schedule of implementation, which encompassed the literature review, planning and data collection with the Company, obtaining Laboratory data, adequacy of the processing, evaluation if the model used by the establishment to comply with the precepts of sustainability, adequacy of proposals sought by the Institute to the template management, preparation of technical material for consultation and training. The samples of water and soil from surrounding marine farms and IVF-Lab were collected by the Institute's technicians and immediately sent for examination in reference laboratory. The laboratory results were presented to the author for verification and inclusion in spreadsheets. The evaluations were conducted in three stages, as the methodology used by Santos (2006).

The first stage referred to the activity process in the context of the establishment concerned, the definition of the impacts scope, the creation and characterization of indicators, according to the characteristics of the activity and the local environment, importance and weight of components and the range of occurrence in the establishment and their surroundings. A first visit to the laboratory and into two Marine Farms in the surroundings was then carried out.

The second consisted of survey and interview with the responsible Company, filling of the array weight of the system to generate partial indexes and aggregates of impact to graphic expression. In the second visit to the Institute took place a meeting with sea farmers and people connected to the activity promotion in Angra dos Reis; the Executive Director of the Institute was interviewed and set the period to be considered for the application of evaluation platform. The third step, immediately afterwards, was referring to the analysis and interpretation of these indexes. The verification of compliance, with an indication of alternatives for correction of non-compliance, in the forms of management and technologies used will be the subject of a work for hiring at the Institute, after presenting the report, to allow, so minimize the negative impacts, enhance the positive and contribute to sustainable local development.

In this study every aspect was composed of a set of indicators organized into automated weighting matrices, in which the components of indicators were evaluated with coefficients of variation, as personal knowledge of the person responsible for the activities in the laboratory. These indicators serve to recognize, in time, the environmental performance and environmental activities at the Institute; characterized the quality environmental management that must be deployed in the company analyzed. The procedure for evaluation of performance of the activity involved a survey with interview conducted by the user of the system – the author-and applied to the Director of the Institute, responsible for managing the IVF-Lab. The interview addressed the obtaining of the coefficients of variation of components, for each of the indicators of socio-environmental performance of the scallop culture, from March 2004, when it was deployed to technological innovation to IVF of *Coquilles* in the Lab, defined as the temporal cutoff for this evaluation. The impacts of the activities under the specific conditions of management in relation to the coefficient of variation of the component have been standardized as the values expressed in the following: for large increase in the component +3, moderate increase in the +1 component, component unchanged 0, moderate reduction in component -1 and large decrease in the component -3. The Director chose the coefficients of variation of components, in particular reason activity and conditions for private management of the laboratory, according to this temporal Court.

The insertion of these coefficients of variation of the components of environmental indicators goes directly into arrays and sequentially in worksheets of ecological performance dimensions and socio-environmental Performance resulted in automatic expression of activity impact index weighted by scale factors of the occurrence and weight of components. The final results of the evaluation of the performance were expressed graphically in the "worksheet, activity, performance" presented at the end of this work, referring to the search results. Automatic arrays included also two weighting factors that relate to the scale of the event and to the weight of the component to the formation of the indicator. The scale of occurrence spells out the space in which the impact of the activity, depending on the particular situation of local application, and can be: punctual when the impact of the activity in the component is restricted to the area or enclosure in which it is occurring the change in the component; local, when the impact make it feel externally to this area, but confined to the limits of the productive unit or establishment; or in the surroundings when the impact reaches beyond the boundaries of the productive unit. The weighting factor of scale implied occurrence in multiplying the coefficient of variation of the performance component of the activity by a predetermined amount, as its scope, namely: punctual 1, local 2 and surroundings 5.

A second weighting factor included in the impact assessment matrices was the weight of the component to the formation of the performance indicator of the activity. The values of the weights of components expressed in the arrays can be changed by the user of the system, to better reflect specific situations evaluation, in which you want to emphasize some of the components, provided that the total weight of the components for a given indicator is equal to the unit (+/-1). The indicators considered in whole defined the overall Performance of the activity, which involved the weighting of the importance of the indicator and the relative weights to the indicators. The composition of this index involves new weighting of the importance of indicators, and the relative weights to the indicators can be changed by the user of the system, provided that the total is equal to the unit (1). With this set of weighting factors, standardized scale in the system varied between -15 and +15, standardized for all indicators individually and for the overall performance of the activity.

The survey used a set of spreadsheets, in MS-Excel, platform that integrated twenty-four indicators of performance for the scallops' productive economic activity at the IEDBIG. A total of 125 components, who understood the variables checked in accordance with their respective weightings amendment. These indicators were grouped into seven aspects and two dimensions, namely: environmental performance and socio-environmental Performance. The indicators were constructed in weighting matrices, in which quantitative data, obtained in the laboratory, were automatically transformed into impact indices, expressed graphically. Should be taken into consideration and emphasized that the report of the results obtained is directed to the producer, where the Director of the Institute, primarily with the indication of management strategies to minimize undesirable impacts and maximize those that contribute to the sustainable performance of the activities in the company, in the lab and surrounding marine farms. Finally, the making of report for the company, article for presentation in scientific events, and publications in specialized journals.

The base system of eco-certification of rural activities, used in this study, was described by Rodrigues et al. (2006) and presented a final worksheet with the results of each aspect studied, after the evaluation of the coefficients of variation for the components of social and environmental impact indicators and the calculation of their weighting in indexes arrays. All indicators were described and the results presented in graphics originating in their spreadsheets with the weighting tables, some of which are entered in this in the book. The results concerning the impact indices for each aspect were automatically expressed graphically in the performance of the activity sheet. Was assessed a set of twenty-four-environmental performance indicators, pectinicultura activity in the IVF Lab and Larvicultura of Scallops, which encompassed a total of 125 components, with verified variables according to their respective weightings amendment, on the dimensions of the ecological and environmental performance, bringing together the 24 arrays of weighting of indicators discussed.

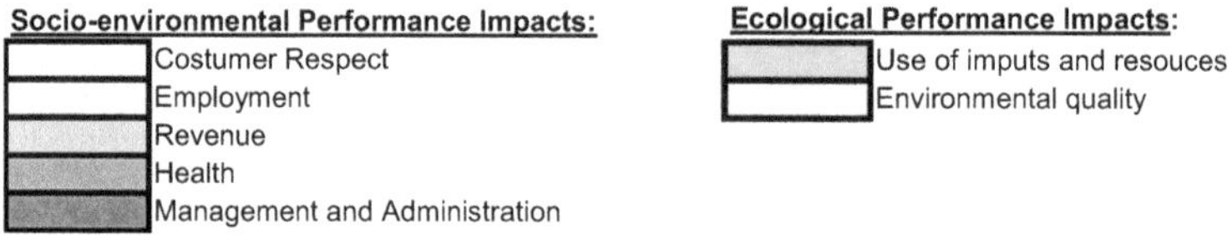

Socio-environmental Performance Impacts:
Costumer Respect
Employment
Revenue
Health
Management and Administration

Ecological Performance Impacts:
Use of imputs and resouces
Environmental quality

Fig. 6. Socio-environmental health quality and landscape ecology: indicators set for the eco-certification of agribusiness activities evaluation system. Data based on Santos (2006).

3.3 Results

The system's spreadsheets automatically calculated impact indices, in a scale ranging between ±15, for 24 indicators from seven essential aspects, were as follows: Use of Inputs and Resources (-2.09), Environmental Quality (3.54), Costumer Respect (5.05), Employment (4.40), Revenue (2.17), Health (3.63), and Management and Administration (1.02). The general socio-environmental performance index (5.04) for the pectens production activities indicated an important contribution of technological innovations for the sustainability of Pectens FIV-Lab production. These results (Figure 7) show the good performance of the proposed system for the evaluation of technology management, agribusiness activities, regional development, and for sustainability performance assessment for pectens production.

The author observed also some improvement related with the management and sustainability of the population who work and live around the Institute: the goal of combating poverty as its line of strategy within the context of its social responsibility, and its main objective to encourage cultivation of endangered species in marine farms, while preserving traditional fishing methods, promoted the creation of more marine farms and worked towards the principle of sustainability and autonomy for low-income fishermen communities based in the region. That means, for example, the reproduction control of fish spat, spawn and fry and research to increase aquaculture productivity. As a result, marine fauna and its biota are being preserved and managed by the technicians themselves, who improve their incomes with the cultivation of endangered species. In addition to giving a tenfold return on capital invested, marine-culture provides jobs and other collateral activities.

Socio-environmental Performance Impacts:

5.05	Costumer Respect
4.40	Employment
2.17	Revenue
3.63	Health
1.02	Management and Administration

Ecological Performance Impacts:

-2.09	Use of imputs and resouces
3.54	Environmental quality

Activity performance index	Indicator weight	Perform coeffic
Use agric. imputs & resources	0.05	0.5
Agric. imputs & raw material	0.05	-3.0
Use of energy	0.05	-12.0
Atmosphere	0.02	1.0
Soil quality	0.05	7.5
Water quality	0.05	-1.0
Biodiversity	0.05	6.0
Environmental rescue	0.05	10.2
Product quality	0.05	3.8
Productive ethics	0.05	9.3
Training program	0.05	10.0
Loc.qualif. Employm.opport.	0.02	6.7
Offer employm. & work cond.	0.05	6.7
Employment quality	0.05	10.8
Gen. income of the enterprise	0.05	15.0
Divers. of Income generation	0.025	7.5
Property value	0.025	9.8
Environm. & personal health	0.02	1.6
Security & professional health	0.02	-1.5
Feeding security	0.05	9.6
Manager & dedication	0.05	11.3
Business & market conditions	0.05	4.5
Waste & water recycle cond.	0.05	-3.0
Institutional relationships	0.02	7.5
Weighted average 1	Activity performance index	5.04

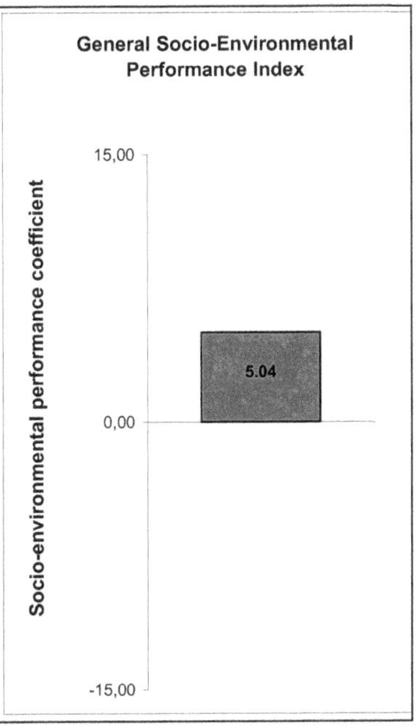

General Socio-Environmental Performance Index

Fig. 7. Results obtained with the socio-environmental health and sustainability assessment for technology innovations at pectens production in Brazil: general socio-environmental indicators and its performance index for the activities.

3.4 Discussion and conclusions

While environmental conservation and social responsibility issues gain increasing importance in the development agendas at all institutional levels, it becomes necessary to select, adapt, transfer, and assess sustainable environmental management and best production practices (Barnthouse et al., 1998). Especial reference to this managerial movement is warranted when agribusiness activities are regarded, because of the spatial scale and bulk of natural and human resources encompassed worldwide by agriculture (Pimentel et al., 1992). In order to bring a practical reach to this sustainable development objective, society should value and recompense farmers and producers who adequately manage their environment and resources, both as an incentive towards sustainability and as repayment for environmental and social services rendered (Viglizzo et al., 2001).

In effect, a multi-attribute EIA system for agribusiness activities' environmental management, integrating dimensions related to landscape ecology, environmental health quality, socio cultural values, economic values, and management values should be applied. It is currently under extensive field application, like this work, with contribution to the stepwise process of sustainable agricultural technology development and appraisal (Rodrigues & Viñas, 2007; Santos, 2010).

Fig. 8. More socio-environmental health quality: aquaculture, plastic lantern and handcraft products courses for the communities.

The IEDBIG established the goal of combating poverty as its line of strategy within the context of its social responsibility (Figure 8), and its main objective is to encourage raising endangered species in marine farms, while preserving traditional fishing methods. The Ilha Grande Bay Marine-Farming Program, known as POMAR, promotes the creation of marine farms and work towards the principle of sustainability and autonomy for low-income fishermen communities based in Angra dos Reis, State of Rio de Janeiro – Brazil (Santos & Zaganelli, 2007). Considering the pectens production and the marine farms activities, 24 socio-environmental indicators were developed and the impact indices were automatically calculated by the system's spreadsheets. General performance index for the pectens reproduction activities indicated an important contribution of technological innovations for the sustainable production of the FIV-Lab. The employed method was considered as appropriate for evaluations of sustainability at this agribusiness activity, dealing with indicators as tools in order to identify possible risks for negative impacts. Those indicators include aspects beyond those commonly presented by environmental impact assessments (Girardin et al., 1999; Bisset, 2000, Rodrigues et al., 2006), and were capable to provide adequate management and sustainable development for the studied Organization. The present paper revised the impact assessment systems, discusses their application for the pectens production with technological innovation management, and aim to the conclusions:

a. The main contributions were to improve the understanding of producers and researchers alike, about the social and environmental implications of the adoption of agribusiness technology innovation, and to introduce socio-environmental impact assessments at an operational level, facilitating the grasp of interactions between technology adoption and the sustainable development of aquaculture.

b. The acceptance of simpler assessment system is an important step toward more complex methods that require a stronger analytical basis and involve a more complex theoretical foundation.

c. The Best SES consists of a practical EIA system for technology innovations, ready for field application through interviews, with directed survey at the FIV-Lab, marine farms and manager responsible for the agribusiness activity, modified by the adoption of the studied technology.

d. The system relies on a computational platform readily available and easily applicable at low cost, and facilitates the storage and communication of information regarding environmental impacts.

e. The computational structure for the system is simple and transparent, unveiling to the user all operations performed with the data. Also, while fairly standardized relative to measurements, the system is malleable, allowing the user to adapt for specific use situations, by changing the weighing factors of indicators and components when appropriate.

This chapter presented and discussed an example of a practical impact assessment method to improve environmental health and we finalize saying: We do believe that in each living creature there is the desire to have a better life, of wealth and contribution, to be happy and make a difference! We can even doubt our ability to accomplish something grandiose and no matter what are the simplest, but it is true that we do. This potential is in every citizen, just decided consciously by such option and in various circumstances, whether at home, at work, whether in the society in which he lives. We all have the power to decide to live well, having not only a good day, but many great days and always there is the chance to change our attitudes and our world for the better. Always! It's never too late, and let's starts now to be a sustainable citizen.

4. Acknowledgment

The author thanks the ICA-Nit time for the together hard working; the Embrapa Researcher, Ecologist Geraldo S. Rodrigues for the contributions to this data and for advising his MBA Thesis; the President Director of IEDBIG, Engineer José L. Zaganelli for the partnership and previous cooperation. Mr. Alex Canella for the English revision and the Publishing Process Manager for the attention and cooperation. This paper was approved by the Veterinary Medicine Academy from Rio de Janeiro State (AMVERJ), Brazil.

5. References

Andrioli, M & Tellarini, V. 2000. Farm sustainability evaluation: methodology and practice. *Agriculture, Ecosystems and Environment*, Vol. 77, pp. 43-52.

Barnthouse, L; Biddinger, G; Cooper, W; Fava, J; Gillett, J; Holland, M & Yosie, T (Ed).s)). 1998. *Sustainable Environmental Management*. Pellston: Society of Environmental Toxicology and Chemistry, Pellston Series Workshops. 102 p.

Bisset, R. 2000. Methods for environmental impact assessment: a selective survey with case studies, a methodology and terminology of sustainable assessment and its perspectives for rural planning. *Agriculture, Ecosystems and Environment*, Vol. 77, pp. 29-41.

Girardin, P; Bockstaller, C & van der Werf, H. 1999. Indicators: tools to evaluate the environmental impacts of farming systems. *Journal of Sustainable Agriculture*, Vol. 13, No. 4, pp. 5-21.

Monteiro, R.C & Rodrigues, G.S. 2006. A system of integrated indicators for social-environmental assessment and eco-certification in agriculture – Ambitec-Agro. *Journal of Technology Management and Innovation*, Vol. 1, No. 3, pp. 47-59.

Payraudeau, S; Hayo, MG & van der Werf, H. 2004.Environmental impact assessment for a farming region: a review of methods. *Agriculture Ecosysistem and Environment*, Vol. 107, pp. 1-19.

Pimentel, D; Stachow, U; Takacs, DA; Brubaker, HW; Dumas, AR; Meaney, JJ; O'Neil, JAS; Onsi, DE & Corzilius, DB. 1992. Conserving biological diversity in agricultural /forestry systems. *Bio Science*, Vol. 42, pp. 354-362.

PROCISUR (PTR). 2006. In: *Regional Technological Platform*. Access on 02.01.2006, Available from < http://www.procisur.org.uy/online/sustentabilidad_inicial.asp >

Rodrigues, G.S & Viñas, A.M. 2007. An environmental impact assessment system for responsible rural production in Uruguay. *Journal of Technology Management and Innovation*, Vol. 2, No. 1, pp. 42-54.

Rodrigues, G.S; Buschinelli, C.C.A; Rodrigues, I.A; Monteiro, R.C & Viglizzo, E. 2006. Sistema base para eco-certificação de atividades rurais. Jaguariúna: Embrapa Meio Ambiente, In: *Boletim de Pesquisa e Desenvolvimento 37*. 40 p. Access on 04.28.2008, Available from: <*http://www.cnpma.embrapa.br/public pdf21.php3?tipo=bo&id=75* >).

Santos, M.R.C. 2006. Gestão da Inovação Tecnológica e do Desempenho Sustentável das Atividades em Laboratório de Fecundação *In vitro* e Larvicultura de Vieiras (*Nodipecten nodosus*). Monografia (*MBA em Gestão de Negócios Sustentáveis*, Universidade Federal Fluminense-Latec, Niterói: 2006). 95p.

Santos, M.R.C & Zaganelli, J.L. 2007. Socio-environmental and sustainability assessment for regional development in Angra dos Reis, RJ – Brazil. In: *Book of Abstracts*. p.83. Access on 04.29.2008, Available from: <*http://www.uni-tuebingen.de/deutsch-brasilianisches-symposium/Symp2007-ed2.pdf* >

Santos, M.R.C & Rodrigues, G.S. 2008. Socio-environmental and sustainability assessment for technology innovations at pectens production in Brazil. *Journal of Technology Management and Innovation*, Vol. 03, No. 03, pp.123-128.

Santos, M.R.C. 2010. *Desempenho Sustentável em Medicina Veterinária: como entender, medir e relatar*. L.F. Livros Ed., ISBN 978-85-89137-15-7, Rio de Janeiro, 186 p.

Viglizzo, E.F; Lértora, F.A; Pordomingo, A.J; Bernardos, J; Roberto, Z.E & Del Valle, H. 2001. Ecological lessons and applications from one century of low external-input farming in the pampas of Argentina. *Agriculture, Ecosystems & Environment*, Vol. 81, pp. 65-81.

Permissions

The contributors of this book come from diverse backgrounds, making this book a truly international effort. This book will bring forth new frontiers with its revolutionizing research information and detailed analysis of the nascent developments around the world.

We would like to thank Dr. Chaouki Ghenai, PhD, for lending his expertise to make the book truly unique. He has played a crucial role in the development of this book. Without his invaluable contribution this book wouldn't have been possible. He has made vital efforts to compile up to date information on the varied aspects of this subject to make this book a valuable addition to the collection of many professionals and students.

This book was conceptualized with the vision of imparting up-to-date information and advanced data in this field. To ensure the same, a matchless editorial board was set up. Every individual on the board went through rigorous rounds of assessment to prove their worth. After which they invested a large part of their time researching and compiling the most relevant data for our readers. Conferences and sessions were held from time to time between the editorial board and the contributing authors to present the data in the most comprehensible form. The editorial team has worked tirelessly to provide valuable and valid information to help people across the globe.

Every chapter published in this book has been scrutinized by our experts. Their significance has been extensively debated. The topics covered herein carry significant findings which will fuel the growth of the discipline. They may even be implemented as practical applications or may be referred to as a beginning point for another development. Chapters in this book were first published by InTech; hereby published with permission under the Creative Commons Attribution License or equivalent.

The editorial board has been involved in producing this book since its inception. They have spent rigorous hours researching and exploring the diverse topics which have resulted in the successful publishing of this book. They have passed on their knowledge of decades through this book. To expedite this challenging task, the publisher supported the team at every step. A small team of assistant editors was also appointed to further simplify the editing procedure and attain best results for the readers.

Our editorial team has been hand-picked from every corner of the world. Their multi-ethnicity adds dynamic inputs to the discussions which result in innovative outcomes. These outcomes are then further discussed with the researchers and contributors who give their valuable feedback and opinion regarding the same. The feedback is then collaborated with the researches and they are edited in a comprehensive manner to aid the understanding of the subject.

Apart from the editorial board, the designing team has also invested a significant amount of their time in understanding the subject and creating the most relevant covers. They scrutinized every image to scout for the most suitable representation of the subject and create an appropriate cover for the book.

The publishing team has been involved in this book since its early stages. They were actively engaged in every process, be it collecting the data, connecting with the contributors or procuring relevant information. The team has been an ardent support to the editorial, designing and production team. Their endless efforts to recruit the best for this project, has resulted in the accomplishment of this book. They are a veteran in the field of academics and their pool of knowledge is as vast as their experience in printing. Their expertise and guidance has proved useful at every step. Their uncompromising quality standards have made this book an exceptional effort. Their encouragement from time to time has been an inspiration for everyone.

The publisher and the editorial board hope that this book will prove to be a valuable piece of knowledge for researchers, students, practitioners and scholars across the globe.

List of Contributors

Chaouki Ghenai
Ocean and Mechanical Engineering Department, Florida Atlantic University, USA

Aniceto Zaragoza Ramírez
Polytechnic University of Madrid, Spain

César Bartolomé Muñoz
Spanish Cement Association, Spain

Armin Grunwald
Institute for Technology Assessment and Systems Analysis (ITAS), Karlsruhe Institute of Technology (KIT), Karlsruhe, Germany

Chaouki Ghenai
Ocean and Mechanical Engineering Department, Florida Atlantic University, USA

M.I. Brasileiro and A.W.B. Rodrigues
Federal University of Ceará, Campus Cariri, Cidade Universitária, Juazeiro do Norte, Brazil

R.R. Menezes
Federal University of Paraíba, UFPB/PB, Cidade Universitária, João Pessoa, Brazil

G.A. Neves and L.N.L. Santana
Federal University of Campina Grande, Campina Grande, Brazil

K.V. Ragnarsdóttir
Faculty of Earth Sciences, University of Iceland, Reykjavik, Iceland

H.U. Sverdrup and D. Koca
Applied Systems Analysis & System Dynamics Group, Department of Chemical Engineering, Lund University, Lund, Sweden

V. Chipofya
University of Malawi, The Polytechnic, Malawi

S. Kainja and S. Bota
Malawi Water Partnership Secretariat, c/o Malawi Polytechnic, Malawi

Edmundo García-Moya and Angélica Romero-Manzanares
Colegio de Postgraduados, México

Rafael Monroy and Columba Monroy-Ortiz
Universidad Autónoma del Estado de Morelos, México

H. Ramesh and A. Mahesha
Dept. of Applied Mechanics & Hydraulics, National Institute of Technology Karnataka, Surathkal, Mangalore, India

Giani Gradinaru
Bucharest Academy of Economic Studies, Romania

Márcio Ricardo Costa dos Santos
Universidade Federal Fluminense (ASPI/UFF), Academia de Medicina Veterinária (AM-VERJ), Instituto Camboinhas Ambiental (ICA-Nit), Brazil

Printed in the USA
CPSIA information can be obtained
at www.ICGtesting.com
JSHW011448221024
72173JS00004B/992